TELEVISION TECHNOLOGY: Fundamentals and Future Prospects

The Artech House Telecommunication Library

TELEVISION TECHNOLOGY:
Fundamentals and Future Prospects

A. Michael Noll

Annenberg School of Communications
University of Southern California

ARTECH HOUSE

Library of Congress Cataloging-in-Publication Data

Noll, A. Michael.
    Television technology.

    Bibliography: p.
    Includes index.
    1. Television.   I. Title.
TK6630.N64   1988   621.388   88-7377
ISBN 0-89006-332-X

International Standard Book Number: 0-89006-332-X
Library of Congress Catalog Card Number: 88-7377

10 9 8 7 6 5 4 3 2 1

Contents

Preface

This book explains the technical workings of television. Conventional broadcast monochrome (black-and-white) and color television are covered along with such alternative delivery systems as video cassette recorders, cable television, and satellites. We also describe new technologies like solid-state cameras, video discs, flat displays, high-definition television, and teletext.

The material presented in the book is written for people with little technical or engineering knowledge. The material is intended to demystify the technical principles of television technology. Such demystification is necessary for managers and other nonengineering people working in the television industry, in broadcasting, production, or consumer electronics. An understanding of the technical workings of television can indeed be obtained by most nontechnical people and will result in improved communication with engineers and other technical personnel.

This book resulted from graduate-level courses in communication technology taught at the Annenberg School of Communications at the University of Southern California and also the Interactive Telecommunications Program at New York University. The material given here formed about a one-semester graduate-level course.

A companion volume on the basic principles of modern electronics (Artech House, 1988) would be useful reading to help understand some of the background technology for television. Such topics as amplitude modulation, frequency modulation, and digital signals are explained in this basic text.

The figures were drawn at the Annenberg School by Richard Cook on a Macintosh SE computer using Cricket Draw software. The professional quality of his work is very much appreciated.

This book on television is the third in a series written by this author and published by Artech House. The other two books in the series treat the telephone system (*Introduction to Telephones and Telephone Systems*, 1986) and basic telecommunication electronics (*Introduction to Telecommunication Electronics*, 1988). The foresight of Artech's former acquisition editor, Barbara Modelski, and Artech's publisher, William M. Bazzy, is most appreciated, since without their vision of a series of introductory books on communication electronics and systems these books would not have been published. Artech's editor of this series has been Dennis Ricci, and he has been a source of much help and improvement in the clarity of the texts. The support and encouragement throughout the writing of the books from the Dean of the Annenberg School, Dr. Peter Clarke, is likewise greatly appreciated.

For
John R. Pierce and Edward E. David, Jr.,
with appreciation for introducing
me and many others
to the joys and wonders of
communication research

1. Introduction

This book is about the technical workings of television, which is perhaps the most exciting form of electronic home entertainment imaginable! Television is so much a part of our lives that it is routinely taken for granted by most of us.

The average person in the United States spends over four hours in front of the television set each day, and the average household television usage per day is more than seven hours. A majority of people now say that television is their main source of news. Also, television is judged to be the most credible source of news as compared to newspapers, radio, and magazines. People obtain their information about the world from television to such an extent that a popular novel, *Being There*, by Jerzy Kosinski, was based on the scenario of a person whose sole source of information about life and the world was television.

The sale of television receivers is the single largest area of consumer electronics. In 1985, the total number of black-and-white receivers shipped to dealers was 3,745,000 units with a retail value of $367 million. The total number of color receivers shipped in that same year was 16,894,000 units with a retail value of $7.25 billion.

From a number of perspectives, television has clearly become the medium with the greatest effect on our lives and on the consumer electronics industry.

Television is considered by many people to be a stable technology without opportunities for further innovation. This belief is far from the truth. Major innovations are being made in television technology in such areas as small portable receivers, improved picture tubes, stereophonic sound, and higher quality circuitry, and the future may bring high-definition television (HDTV), large flat viewing screens, and digital circuitry. Television is thus an exciting area of continuing technological innovation.

Television is a sophisticated electronic system that represents the ultimate of analog technology. An understanding of the technical workings of television is not beyond the grasp of educated nontechnical people, and this book attempts to develop that understanding.

We start with a review of the history of television's invention and development in Chapter 2. The basic ideas and concept of television are actually quite old, predating the twentieth century. The basic workings of television are described in Chapter 3, in terms of the signals used to convey the picture information. These signals combine information about the brightness variations in the scene, along with information necessary to synchronize the receiver in the home with the signals being transmitted by the television station. The television signal requires a very large amount of *bandwidth*. Modulation, bandwidth, and waveforms are discussed in Chapter 4.

Color television is described in Chapters 5 and 6, in terms of both the signals themselves and their spectral implications. Color television was able to transmit the extra information for color and yet maintain compatibility with existing monochrome television, which was truly an ingenious innovation. The National Television System Committee (NTSC) standard for color television is used in the United States, Japan, and other countries. To understand color television, it is necessary to understand color theory, and so our discussion of color television is introduced by a review of *colorimetry*.

Picture tubes and camera tubes are essential for the workings of television, and these two technologies are described next in Chapter 7. Monochrome and color receiver circuitry are then explained along with the prospects for digital circuitry. Digital circuitry allows a number of new features in the television receiver as well as the ability to improve the quality of the picture.

Broadcast television is only one of the ways in which television programs are received in the home (Chapter 8). The *video cassette recorder* (VCR) is one of the alternatives to broadcast television, and the VCR allows shows to be taped for later viewing in addition to watching movies rented from neighborhood stores. The VCR uses some innovative approaches to record the wide bandwidth needed for the television signal. Another alternative for distributing television programs was the video disc, which failed in the marketplace because of the recording ability of the VCR. VCRs and video discs are described in Chapters 9 and 10.

Over half the television households in the United States receive television programs over a coaxial cable network operated by a local cable television (CATV) company. CATV and other alternative broadcast technology such as direct broadcast satellites, low-power television (LPTV), and subscription television are explained in Chapter 11. Optical fiber is being considered as the medium of the future for delivering telephone and television signals to the home. This technology is described together with some of the policy issues that such a combination of these two services will create.

Chapters 12 to 16 describe present and forthcoming innovations in television technology. Stereophonic sound is now being transmitted by many television stations, and we explain how this transmission is achieved. New display technologies such as projection television, flat displays, and liquid crystal displays (LCD) are described. Solid-state, charge-coupled-device, image sensors and their use for color cameras, new uses of the vertical blanking interval portion of the television signal to transmit digital information, and the prospects for improved higher-definition television with a discussion of the standards issue are covered.

The last chapters cover international television standards (Chapter 17) and present an overview of the future of television technology (Chapter 18). The PAL system for color television is used in the United Kingdom and other countries, while the SECAM system is used in France and other countries.

A glossary of some of the technical terms used in the text is presented at the end of this book.

Television is truly a marvelous invention and an exciting technology! This book hopes to accomplish a demystification of the technical workings of television so that more people can share in its excitement and continuing innovation.

References

'85 Nielsen Report on Television, A. C. Nielsen Company, 1985.

Statistical Abstract of the United States 1987, U.S. Department of Commerce, 1987, p. 751.

2. History

We think of television as being a post–World War II innovation. Actually, the earliest forms of television were first proposed and invented in the 1880s and in the first quarter of the twentieth century. This chapter reviews the early days of television and the pioneers who devoted their lives to making television a reality.

The earliest methods for scanning, dissecting, transmitting, and re-creating images were based on electromechanical technology. One of the earliest patents for such a system was granted to Paul Nipkow of Germany in 1884. The system he disclosed used a rotating disk to scan the scene and to reproduce the scanned image. Each disk contained a spiral array of holes or apertures near the outer edge, as shown in Figure 2-1. A light-sensitive photocell responded to the variations in light in the scanned scene.

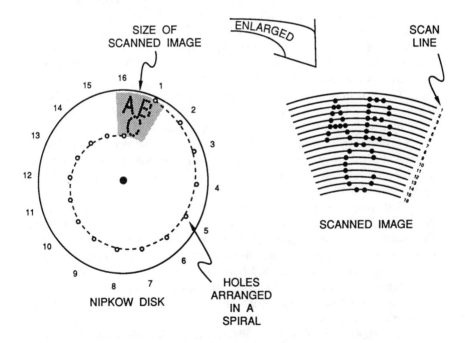

Fig. 2-1 The use of a spiral of holes in a rotating disk to scan an image was invented in 1884 by Paul Nipkow in Germany. The reproduced image was very small with poor resolution.

Synchronized, rapidly rotating prismatic rings and disks were used by Charles Francis Jenkins in a crude television system that was able to show moving images on a receiver screen. His earliest patent on such a television system was filed on March 13, 1922. In 1925, Jenkins used his system to transmit television pictures by radio from Washington, D.C., to Philadelphia. Jenkins continued his research into television using various forms of electromechanical image scanners and displays. Although Jenkins saw the commercial possibilities for television, the Great Depression intervened, and his company, the Jenkins Television Corporation, went bankrupt. The Radio Corporation of America (RCA) acquired rights to many of Jenkins' inventions.

Early experiments in television were conducted in London by John L. Baird around 1925. Baird's experiments were similar to those of Jenkins, except that wire transmission was used. Baird is credited with the first public demonstration in early 1926 of what can be considered true television, in that gradations of light and shade were reportedly visible as opposed to only crude outlines. A few years later, regular television service was broadcast by the British Broadcasting Company (BBC) on an experimental basis using Baird's system. By 1932, about 10,000 television receivers using a Nipkow disk with about 30 lines of resolution had been sold in the United Kingdom.

The first long-distance telecast was made in April 1927 from Washington, D.C., to New York using a purely electromechanical system devised by researchers at Bell Telephone Laboratories. The term "television" first appeared in descriptions of this system by the press. The display consisted of an array of multielectrode neon tubes and a motor-controlled scanning system. Synchronizing signals were transmitted along with the brightness signal.

A display utilizing a projection system with a drum of mirrors was developed in the same time frame by researchers at the General Electric Company. A year later, in 1928, Bell Labs researchers used their electromechanical system with suitable modifications to transmit color images between Washington, D.C., and New York.

Major problems with these electromechanical systems were flicker and poor resolution. These problems were solved by purely electronic scanners and displays.

The application of a purely electronic display, the *cathode ray tube* (CRT), for television was first proposed by Boris Rosing in St. Petersburg, Russia, in 1907. In 1911, he demonstrated a system that produced crude images using a rotating-mirror imager at the transmitter and a cold-cathode picture tube at the receiver. Magnetic coils were used to generate the scanning signals needed for the picture tube.

Vladimir K. Zworykin was a student of Professor Rosing from 1910 to 1912. Zworykin left Russia and came to New York in 1919, and soon was working on television for the Westinghouse Electric and Manufacturing Company in Pittsburgh. In 1923, Zworykin applied for a patent on his ideas, which would later become the basis for modern electronic television. In 1924, while at the Westinghouse Research Laboratories, Zworykin demonstrated a television receiver using a seven-inch CRT with electrostatic and electromagnetic deflection of the electron beam. In the same year, he also demonstrated an electronic device, called the *iconoscope*, for electronically scanning a scene.

In 1925, Zworykin filed a patent application for an all-electronic color television system. David Sarnoff, then Executive Vice President of RCA, believed in the commercial possibilities for television and thus increased his support of Zworykin's research at Westinghouse. In late 1929, Zworykin left Westinghouse to direct television research for RCA. Zworykin is credited with being the developer of commercial broadcast television for RCA.

In 1929, working independently, Philo T. Farnsworth demonstrated a completely electronic television system using a CRT with electromagnetic deflection as the display. As early as September 1927, he had developed an all-electronic camera called an *image dissector*. In 1931, he joined Philco and continued his research there.

The first regular schedule of television broadcasting occurred in 1939. The National Broadcasting Company (NBC) stations in New York, Schenectady, and Los Angeles broadcast two one-hour programs per week. On July 1, 1941, the NBC and the Columbia Broadcasting System (CBS) stations in New York were licensed by the Federal Communications Commission (FCC) as the first commercial TV stations in the United States. The television broadcasts used 525 horizontal scan lines, as then standardized by the FCC.

World War II intervened in the growth of television broadcasting. In 1945, the FCC allocated spectrum space for thirteen VHF television channels. Channel 1 was later deleted. In April 1952, the FCC expanded television broadcasting by the allocation of enough UHF spectrum space for an additional seventy new television channels.

Television achieved spectacular market penetrations in a very short time. In 1949, television had a market penetration of 6 percent of American homes. By 1953, this penetration had increased to 49 percent, one of the fastest market penetrations ever seen! (See Schramm, *et al.*, Table II-1.) In its first five years, television achieved an average annual growth rate of 320 percent per year. (See Hough, Table 1.)

References

Abramson, Albert, "Pioneers of Television — Vladimir Kosma-Zworykin," *J. SMPTE,* Vol. 90, July 1981, pp. 580–590.

Abramson, Albert, "Pioneers of Television — Charles Francis Jenkins," *J. SMPTE,* Vol. 95, February 1986, pp. 224–238.

Bingley, F. J., "A Half Century of Television Reception," *Proc.IRE,* Vol. 50, May 1962, pp. 799–805.

Freeman, John P., "The Evolution of High-Definition Television," *J. SMPTE,* May 1984, pp. 492–501.

Gray, Frank, J. W. Horton, and R. C. Mathes, "The Production andUtilization of Television Signals," *Bell System Tech. J.,* Vol. 6, 1927, pp. 560–603.

Hough, Roger W., "Future Data Traffic Volume," *Computer,* September-October 1970, pp. 1–8.

Ingram, Dave, *Video Electronics Technology,* Tab Books, (Blue Ridge Summit, PA), 1983, pp. 17–33.

Reitan, Edwin Howard, Jr., "The Collection of Television Technology and Design," *IEEE Trans. Consum. Electron.,* Vol. CE-30, May 1984, pp. 46–61.

Schramm, Wilbur, Jack Lyle, and Edwin B. Parker, *Television in the Lives of Our Children,* Stanford University Press (Palo Alto, CA), 1961.

3. Basic Principles

IMAGE SCANNING

Television accomplishes its transmission of moving images in a manner quite different from cinematography. In cinematography, a complete two-dimensional image is stored on film and is projected in its totality on the screen for viewing. No such instantaneous two-dimensional representation is possible with television, since only a one-dimensional signal can be transmitted over the air in a serial fashion. This means that two-dimensional imagery must be converted into a serial signal. Such conversion is accomplished by scanning the image with an electronic camera and then reproducing the scanned image at the television receiver, as shown in Figure 3-1.

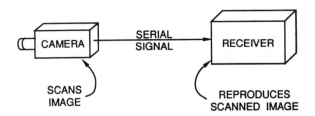

Fig. 3-1 The scene is scanned by an electronic camera that generates a serial signal for transmission to the television receiver. The scanned image is reproduced on the electronic display screen of the television receiver.

With scanning, a number of nearly horizontal passes are made across the image. With each pass across the image, the light reflected by the image is converted into an electrical signal, the instantaneous amplitude of which represents the reflected light energy. Each pass across the image by the scanning element is called a *scan line*.

The television camera at the studio contains an image tube that automatically scans the scene and generates an electrical output, which is a serial signal representing a series of scan lines. The television receiver receives the serial signal and then reproduces, or "writes," the scanned image on a viewing screen. The reproduction is accomplished by a writing, or reproducing, spot that scans across the viewing screen and varies in intensity in accordance with the instantaneous amplitude variation of the received serial signal, as depicted in Figure 3-2.

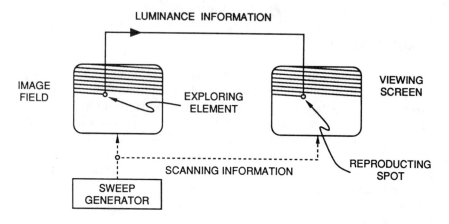

LUMINANCE INFORMATION

IMAGE FIELD

EXPLORING ELEMENT

VIEWING SCREEN

SCANNING INFORMATION

REPRODUCING SPOT

SWEEP GENERATOR

Fig. 3-2 An exploring element is swept horizontally in a scan line across the image field to create a signal corresponding to the variations in the brightness of the scene. A reproducing spot scans across the viewing screen in exact synchrony with the exploring element. After each scan line is completed, the exploring element is moved down vertically a small distance to begin the next scan line.

Each scan line usually is horizontal, and the scanning starts at the top of the image. The scan line is then moved down vertically a small distance so that the next scan across the image can be performed. The electrical signal that causes the horizontal scans to occur is called the *horizontal sweep signal,* since the scanning element "sweeps" horizontally across the image. The electrical signal that moves each horizontal scan line vertically is called the *vertical sweep signal.*

Although not used for broadcast television, each scan line could be perfectly horizontal, as shown in Figure 3-3. The horizontal sweep signal would be a ramp with a positive slope corresponding to moving the scan from left to right. A much shorter ramp with a negative slope would move the scan back to the leftmost position to begin the next scan. This is called the *retrace interval.* The vertical sweep signal would be constant during each scan and would then be a short ramp to move the scan down to the next line.

The vertical sweep signal actually used in broadcast television continuously moves the scanning element from the top to the bottom at a slow rate compared to that of the horizontal sweep signal. This means that each scan line is slightly tilted to the right. When the scanning element reaches the bottom of the image, it is quickly swept back to the top of the image during the vertical retrace interval. The periodic waveforms

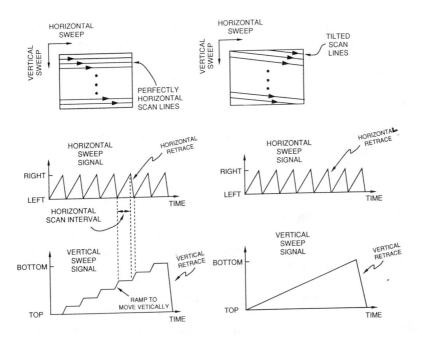

Fig. 3-3 The exploring element is controlled by horizontal and vertical sweep signals. The interval of time during which the exploring element returns to the left side of the image field is called the horizontal retrace interval, and the time during which the element returns to the top of the image field is the vertical retrace interval. The left portion shows scanning of a scene using perfectly horizontal scan lines and the appropriate horizontal and vertical sweep signals. The scan lines actually used in broadcast television are tilted downward slightly to the right, as shown in the right portion. The sweep signals look like the teeth of a saw and are hence called saw-toothed wave-forms.

that control the horizontal and vertical sweeps of the reproducing spot are called *saw-toothed waveforms* because they resemble the teeth of a handsaw.

In television, scanning occurs very rapidly; a complete picture is scanned every thirtieth of a second. Images can be scanned at a much slower rate, as in, for example, facsimile transmission where a single page is scanned in a few minutes for transmission over the telephone network.

SIGNALS

The information about the varying light intensity of the image produced by the television camera is called *luminance information*. The luminance information is transmitted to the television receiver where it is displayed on the viewing screen. The viewing screen is scanned in a manner exactly like the scanning that occurs at the camera. Clearly, the scanning done at the viewing screen must be in exact synchrony with the scanning done at the camera. If not, then the viewing screen might be displaying information at the bottom of the screen when the received luminance signal might contain information about the top of the scanned image.

The scanning at the camera is controlled by the horizontal and vertical sweep signals which are generated by appropriate sweep generators. This scanning information needs to be transmitted to the television receiver so that the motion of the reproducing spot can be synchronized with the appropriate luminance information obtained at the camera. There also is a need to turn off the reproducing spot during the horizontal and vertical retrace intervals, or otherwise its retrace path would be visible on top of the reproduced image. What this means is that in addition to the luminance signal, three other types of information or signals need to be transmitted to the television receiver: one to control the horizontal sweep, another to control the vertical sweep, and one to turn off the reproducing spot during the retrace intervals. All of these signals are combined into one composite signal in broadcast television, as depicted in Figure 3-4.

VERTICAL RESOLUTION

The total number of horizontal scan lines determines the vertical resolution of the scanned image that is reproduced on the viewing screen. The number of scan lines was specified based upon a consideration of the acuity of human vision.

The human eye is able to resolve two parallel lines if the lines subtend about 2 minutes of arc at the eye. Two minutes of arc is 2/60 or 1/30 of

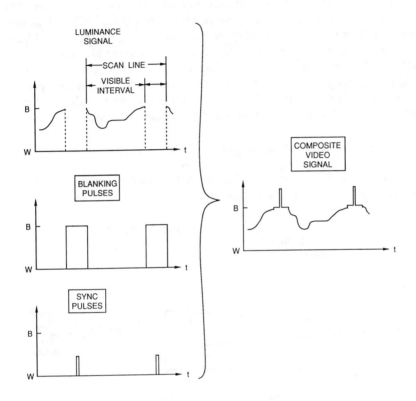

Fig. 3-4 The luminance signal, blanking pulses to extinguish the reproducing spot during the retrace intervals, and the sync pulses used to synchronize the horizontal and vertical sweep signals at the receiver with those at the television studio are all combined into a single signal called the composite video signal.

a degree. The closest viewing distance is assumed to be about four times the height of the picture, or 4*H*. Assume that the distance between the two lines to be resolved is *d*. These two lines should subtend no less than 1/30 of a degree to be resolvable by the human eye. As shown in Figure 3-5, a right triangle is formed with one side equal to the viewing distance, 4*H*, and the other side equal to *d*/2. The angle Θ formed by the hypotenuse and the viewing distance is half the angle subtended by the two parallel lines being resolved, or (1/30)/2 degree. By trigonometry, the tangent of this angle is the opposite side divided by the adjacent side, or

$$\tan(\theta) = \tan(1/60°) = (d/2)/4H = d/8H.$$

This equation can be solved for the number of resolvable lines in a picture of height *H*, or *H*/*d*. The result is about 430 resolvable lines.

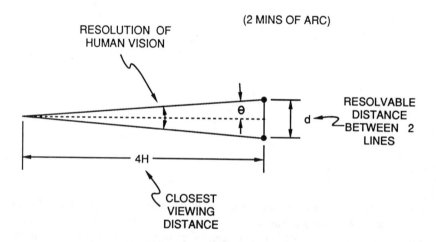

Fig. 3-5 The geometry of the resolvable distance between two horizontal lines in terms of the resolution of human vision.

In commercial broadcast television, the portion of the transmitted image at the very top and very bottom are blocked by the mask around the viewing screen. This is done deliberately, since all television receivers behave differently, and it is expected that a portion of the transmitted image will not be visible on many television receivers. A total of about 53 scan lines must be allowed for this effect, commonly called *overscanning*. This gives a total of about 483 usable scan lines, although as derived in the preceding paragraph only about 430 would be visible on many television receivers because of overscanning.

Time must be allowed for the reproducing spot to retrace itself from the bottom to the top of the viewing screen. This vertical retrace interval is equivalent to the time taken for an additional 42 horizontal scan lines. Thus, the total number of horizontal scan lines in commercial broadcast television is 525 lines.

INTERLACING

A single frame of a television image comprises a total of 525 horizontal scan lines. If these frames were displayed at a rate of 30 frames per second, they would exhibit visual flicker. The frames could be displayed at twice that rate, namely, 60 frames per second, but then the rate at which the information was being transmitted would double, thereby requiring twice as much bandwidth. Images fuse and appear flicker-free for most humans at rates faster than about 40 Hz. Cinema movies are shot at 24 frames per second. If they were displayed at a rate of 24 images per second, the flicker would be objectionable. The trick is to display each frame twice using a shutter rate of 48 images per second. A similar trick is used in television.

Each frame is divided into two fields with one field composed of the odd-numbered horizontal scan lines and the other field composed of the even-numbered horizontal scan lines. Clearly, the field rate is twice the frame rate, or 60 fields per second, thereby avoiding visual flicker. If each field were simply displayed on top of the other field, however, resolution would be halved. The solution is to interlace the two fields so that the scan lines for the odd field fall exactly in between the scan lines for the even field. This technique is called *interlacing* and is depicted in Figure 3-6.

Interlacing preserves the vertical resolution of 525 scan lines and avoids visual flicker of large areas of the reproduced image. Flicker, however, can occur in a small area of the image. For example, if a scene with a horizontal line were scanned, then this line would appear only in alternate fields. It hence would be displayed at a rate of 30 images per second and would flicker. Luckily, most scenes are complex enough that small area flicker is not a problem in commercial broadcast television.

The total number of horizontal scan lines in a frame is 525 lines. Each frame is composed of two fields, and each field consists of 262.5 lines. Field 1 consists of all the odd-numbered lines in a frame, and field 2 consists of all the even-numbered lines. The first line of the odd field starts at the upper-left corner of the picture, and the last (263rd) line of the odd field is a half line that ends at the bottom half of the picture. The first line of the even field begins at the top half of the picture, and the

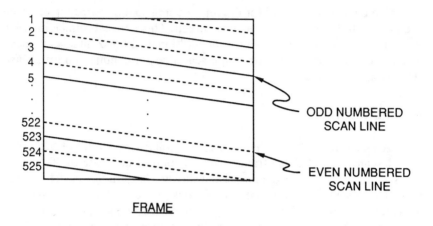

FRAME

Fig. 3-6 An image frame is composed of two fields. The odd-numbered scan lines comprise one field, and the even-numbered lines the other field. A total of 525 scan lines comprise a frame, and each field consists of 262.5 scan lines. The interlacing reduces visible flicker.

last line of the even field ends at the bottom right corner of the picture. The composition of the individual fields is shown in Figure 3-7.

Time is needed for the reproducing spot to travel from the bottom of the screen back to the top to begin the next field. As was mentioned earlier, this time is called the vertical retrace or blanking interval. The amount of time required for 21 scan lines is allowed in each field for the vertical retrace interval. By convention, the *vertical blanking interval* begins each field. Thus, as shown in Figure 3-8, field 1 ends at a half line, and the vertical retrace for field 2 begins immediately. The vertical retrace interval is assigned the time required for 21 lines, and hence the visible portion of field 2 does not start until line 22, as shown in Figure 3-9. The image portions of both fields 1 and 2 begin with line 22.

The aspect ratio of the television picture is 4 to 3, as shown in Figure 3-10. This means that every 4 units in width are matched by 3 units in height. Typically, a portion of the transmitted image at the edges and the top and bottom will not be visible on many television screens. This is because of variations in television receiver circuitry and also because most picture tubes are not perfectly rectangular. Thus, because of the image roll-off, only 90% of the image area is "safe" for most action in terms of being visible on most television receivers. The safe title area is 80% of the total image. The safe areas are depicted in Figure 3-11.

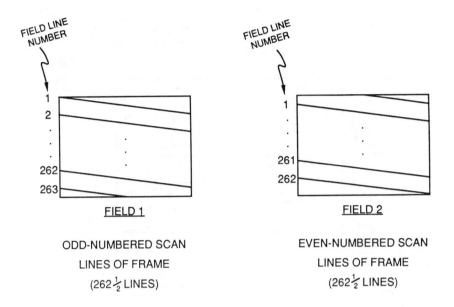

FIELD LINE NUMBER

FIELD 1

ODD-NUMBERED SCAN
LINES OF FRAME
$(262\frac{1}{2}$ LINES)

FIELD 2

EVEN-NUMBERED SCAN
LINES OF FRAME
$(262\frac{1}{2}$ LINES)

Fig. 3-7 The composition of the two fields. The scan lines in each field are individually numbered in sequence. Field 1 ends at the first half of line 263. The remaining portion of this line begins field 2, although line 1 of field 2 is the first full scan line of the field.

HORIZONTAL SCANNING FREQUENCY

The television image is composed of 525 horizontal scan lines per frame. The frames are transmitted at a rate of 30 frames per second. Hence, the horizontal lines occur at a rate of $30 \times 525 = 15,750$ lines per second. The *horizontal scanning frequency* thus is 15,750 Hz.

It will be shown later that the horizontal scanning frequency had to be slightly changed for color television transmission. Thus, the horizontal frequency of 15,750 Hz is only for monochrome television transmission.

WAVEFORMS

The horizontal and vertical sweep signals in the television receiver are generated by oscillator circuits that create the saw-toothed waveforms needed to control the horizontal and vertical movement of the reproducing

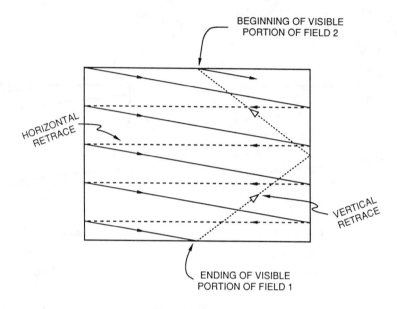

BEGINNING OF VISIBLE
PORTION OF FIELD 2

HORIZONTAL
RETRACE

VERTICAL
RETRACE

ENDING OF VISIBLE
PORTION OF FIELD 1

Fig. 3-8 The visible portion of field 1 ends at the first half of a scan line. Field 2 begins immediately with the vertical retrace interval of exactly 21 lines in duration. The visible portion of field 2 begins at the latter half of the scan line.

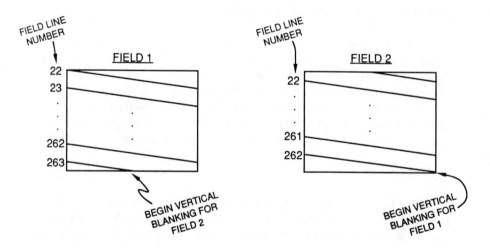

FIELD LINE
NUMBER

FIELD 1

22
23
.
.
262
263

BEGIN VERTICAL
BLANKING FOR
FIELD 2

FIELD LINE
NUMBER

FIELD 2

22
.
.
261
262

BEGIN VERTICAL
BLANKING FOR
FIELD 1

Fig. 3-9 The first 21 lines of each field are reserved for the vertical retrace or blanking interval. The actual image starts at line 22 of each field.

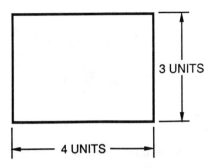

Fig. 3-10 The aspect ratio for television.

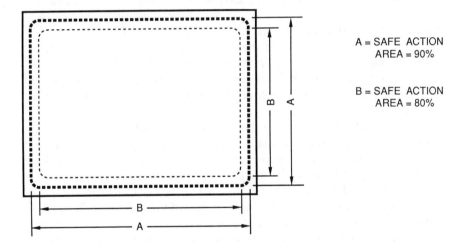

A = SAFE ACTION
AREA = 90%

B = SAFE ACTION
AREA = 80%

Fig. 3-11 The safe action and title areas for a television frame.

spot, as shown in Figure 3-12. The frequency of these oscillators is determined by pulses at their input. For example, the horizontal sweep repeats itself at the horizontal scanning rate of 15,750 Hz. The frequency of the horizontal sweep oscillator is controlled in precise synchrony with the movement of the scanning element at the camera by pulses that are transmitted along with the luminance information. These pulses are called *horizontal synchronization pulses.*

The reproducing spot must be turned off during the horizontal and vertical retrace intervals. This is accomplished by transmitting pulses that extinguish, or blank, the spot during these intervals. These pulses correspond to a black level and are called *blanking pulses.*

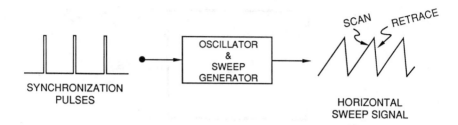

Fig. 3-12 Synchronization pulses control the precise frequency of the horizontal and vertical sweep signals.

In an actual television composite video waveform, the luminance or video waveform, the synchronization pulses, and the blanking pulses are all added together to create a single waveform. The precise shape and timing of this composite waveform is specified by the FCC.

The major component of the composite video waveform is the luminance signal which conveys the information about the varying light intensity in the scanned image. The television waveform is usually plotted in an inverted fashion in which blacker signals are plotted in a positive or upward direction, as shown in Figure 3-13.

The blanking pulses are at the reference black level and the brightest luminance information is at the reference white level. If the tips of the horizontal synchronization pulses are at a level of 100 units, the blanking level is specified to be at 75 units, the reference black level is at 67.5 units, and the reference white level is at 12.5 units. The tolerance on all these levels is specified as ± 2.5 units. A plot of the waveform for a single horizontal scan line is shown in Figure 3-14.

For a horizontal scanning frequency of 15,750 Hz, the spacing between the horizontal synchronization pulses is 1/(15,750) second, or 63.4 microseconds (abbreviated μs). The spacing between the horizontal synchronization pulses, or horizontal "sync" pulses as they are usually called, is symbolized as H. Although, the timings of the various components of the television waveforms are specified by the FCC in terms of H, typical timings are as follows. The horizontal sync pulses have a width of 5.1 μs and are situated on top of the horizontal blanking pulses. The horizontal blanking pulses are about 10.2 μs wide. The horizontal sync pulse begins 1.3 μs after the beginning of the horizontal blanking pulse, and this interval is called the *front porch*. There is an interval of 3.8 μs after the horizontal sync pulse before the horizontal blanking pulse ends; this interval is called the *back porch*.

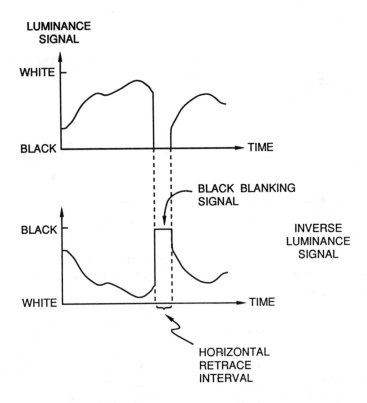

Fig. 3-13 The luminance signal is usually depicted as an inverse video signal for which black is plotted in the up or positive direction.

The vertical blanking interval occurs at the beginning of each field to turn off the reproducing spot while it retraces from the bottom of the picture back to the top to begin scanning the next field. A total of 21 horizontal scan lines are allowed for the vertical retrace interval. The vertical blanking pulse exists for this interval. The vertical blanking pulse is depicted in Figure 3-15.

The vertical sweep oscillator is controlled by the vertical sync pulses. A vertical sync pulse occurs at the end of each field along with the vertical blanking pulse. A vertical sync pulse has a width equal to three times the horizontal line width of H seconds. The vertical sync pulse has five serrations inserted at intervals of $H/2$ seconds. These serrations assure horizontal synchronization during the vertical sync pulse. Each serration is 4.4 μs wide.

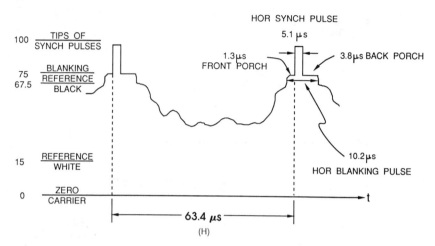

Fig. 3-14 The waveform of a single horizontal scan line. For monochrome television, each horizontal scan line is 63.4 μs in length. For color television, each scan line is 63.5 μs long.

The vertical sync pulse is immediately preceded and followed by portions of the vertical blanking interval equal to $3H$. Six equalizing pulses spaced at $H/2$ seconds occur during each of these two intervals. These equalizing pulses assure horizontal synchronization and minimize interlace problems. Thus, the total width of the vertical sync pulse and the preceding and following series of equalization pulses is $9H$. This sequence is then followed by the remainder of the vertical blanking pulse, during which time the horizontal sync pulses continue to occur at their normal spacing.

The vertical blanking pulse along with the vertical sync pulse is initiated at half the 263rd line of field 1 and at the end of the 262nd full line of field 2. Thus, there is a half line difference in the beginning of the vertical blanking pulse for every other field. It is for this reason that the equalizing pulses must be placed every $H/2$ seconds.

PROGRESSIVE SCANNING

Interlaced scanning is used with conventional broadcast television to prevent flicker and also to present good vertical resolution without requiring an excessive number of scan lines. Each full frame of 525 scan lines is divided into two fields of 262.5 scan lines each, and although the frame rate is only 30 frames per second, the field rate of 60 fields per second is sufficient to prevent visible flicker.

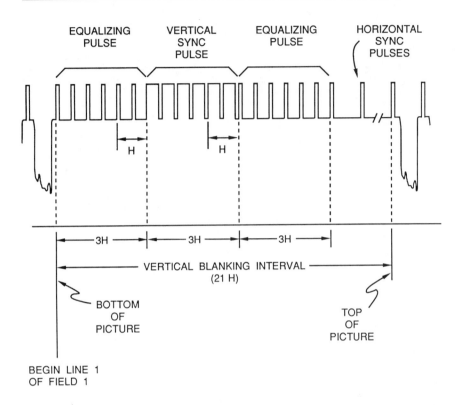

Fig. 3-15 The waveform of the vertical sync pulse and remainder of the vertical blanking pulse. Equalizing pulses assure that horizontal synchronization will be maintained, even though alternate fields begin on half lines.

An alternative approach, used in some industrial television systems, is to avoid interlacing altogether. These noninterlaced systems transmit the 525 scan lines progressively in each full frame, and there are no half-frame fields. The frame rate must then be increased to 60 frames per second to prevent visible flicker. Vertical resolution is improved, but the bandwidth required to transmit the signal is doubled compared with conventional interlaced television.

References

Fink, Donald G., "Television Fundamentals and Standards," in *Electronics Engineers' Handbook (Second Edition)*, Donald G. Fink (Editor-in-Chief), McGraw-Hill (New York), 1982, pp. 20-3 to 20-23.

4. Monochrome Television Transmission

MODULATION

The composite video information is transmitted separately from the audio information in television transmission. The composite video information is transmitted as vestigial amplitude modulation of a radio-frequency (RF) carrier. The audio information is transmitted as frequency modulation of a separate RF carrier. These modulation techniques and their bandwidth implications are described in this chapter.

VIDEO BANDWIDTH

The bandwidth required for the composite video signal is determined by the maximum frequency component that is present in the luminance signal. This maximum frequency component can be estimated by an analysis of the resolution in each horizontal scan line.

Clearly, the horizontal resolution should match the vertical resolution which is determined by the number of scan lines. A total of 525 scan lines forms each frame. Of these 525 scan lines, 42 are used for the vertical retrace, leaving a total of 483 usable visible lines. If a horizontal line in the image being scanned were to fall exactly between two adjacent scan lines, it would not be shown well. The empirically determined relationship between the number of visually resolvable lines and the number of scan lines is called the Kell factor and is about 70%. This effect results in a decrease in vertical resolution by about 70% so that the number of resolvable vertical lines is $0.70 \times 483 = 338$ lines. The resolution in the horizontal direction should be comparable to the vertical resolution. Hence, the vertical resolution of 338 lines needs to be multiplied by the aspect ratio of 4/3 thereby giving $338 \times 4/3 = 451$ resolvable points as the horizontal resolution comparable to the vertical resolution.

The worst case for horizontal resolution would be a series of closely spaced vertical bars. The vertical bars when scanned would generate a luminance signal that rapidly alternated between white and black. The fastest rate of this alternation that would be resolvable would be $451/2 = 225$ cycles. This number of cycles occurs during the visible portion of the horizontal scan line. The visible portion of a horizontal scan line equals 63.4 μs minus the 10.2 μs required for the horizontal retrace interval, or $63.4 - 10.2 = 53.2$ μs. A total of 225 cycles occurring in 53.2 μs is comparable to a frequency of 4.2 million cycles per second, or 4.2 MHz.

Thus, if the luminance signal is band-limited to 4.2 MHz, it will have a horizontal resolution comparable to the vertical resolution that is determined by the number of scan lines per frame. Actually, the FCC allows a bandwidth of 4.2 MHz for the video signal, so the preceding analysis is indeed a reasonable justification of this allowed bandwidth.

With such a large bandwidth for the video signal, the only type of modulation that would be reasonable for the video signal is amplitude modulation. The result of amplitude modulation is to move the spectrum of the modulating signal, also called the baseband signal, to a higher frequency region in the spectrum. Usually, the modulated carrier has a spectrum consisting of two components: an upper sideband and a lower sideband. The upper sideband is an exact replica of the spectrum of the baseband signal shifted to a spectral region above the carrier frequency. The lower sideband is a mirror image of the spectrum of the baseband signal shifted to a spectral region below the carrier frequency. Thus, if the bandwidth of the baseband signal is W, the bandwidth of the amplitude modulated carrier is the sum of the bandwidths of the upper and lower sidebands, or $2W$. This type of amplitude modulation is called *double-sideband* (DSB) modulation.

Clearly, if conventional double-sideband amplitude modulation were used for the video signal, a bandwidth of $2 \times 4.2 = 8.4$ MHz would be required. This is a very large amount of radio spectrum space. A more spectral-efficient scheme is used for television transmission.

The lower sideband is an exact replica of the upper sideband and hence contains redundant information. It can thus be eliminated resulting in *single-sideband* (SSB) transmission. The problem with single-sideband transmission is that the receiver is somewhat complex. A solution is to transmit only a portion, or a "vestige," of the lower sideband. This is called *vestigial amplitude-modulation transmission* and is the method used for commercial broadcast television. The vestigial lower sideband has a nominal or minimum width of 0.75 MHz and is required to be no lower than 1.25 MHz below the frequency of the carrier.

AUDIO BANDWIDTH

The audio signal is transmitted as frequency modulation of an RF carrier situated 4.5 MHz above the frequency of the video RF carrier. Frequency modulation usually consumes more bandwidth than amplitude modulation but is considerably less prone to noise.

The spectrum of the frequency-modulated carrier extends from 50 kHz below the carrier frequency to 50 kHz above the carrier frequency for a total bandwidth of 100 kHz. The baseband audio signal is limited to audio frequencies from 50 Hz to 15 kHz.

An important parameter for frequency modulation is the maximally allowable swing, or maximum deviation, in the carrier frequency as the modulating signal reaches its maximum amplitude. The maximum frequency deviation is an important factor in determining the bandwidth of a frequency-modulated carrier. For television audio, the maximum frequency deviation is specified by the FCC as ±25 kHz.

CHANNEL BANDWIDTH

The bandwidth occupied by a broadcast television channel consists of the bands required for the composite video signal and the audio signal, as shown in Figure 4-1. The composite video occupies a spectrum from 1.25 MHz below the video carrier to a little less than 4.5 MHz above the video carrier. The upper sideband for the audio signal extends 50 kHz above the audio carrier of 4.5 MHz. A frequency space of 200 kHz is left as a guard band between adjacent television channels.

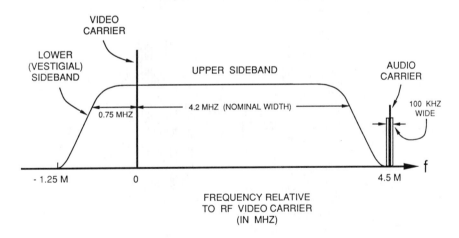

FREQUENCY RELATIVE
TO RF VIDEO CARRIER
(IN MHZ)

Fig. 4-1 The bandwidth occupied by a monochrome television channel extends from 1.25 MHz below the video radio-frequency carrier to about 4.75 MHz above the video carrier for a total bandwidth of 6.0 MHz. The video signal is restricted to frequencies less than 4.2 MHz. The audio signal is transmitted on a separate radio-frequency carrier located 4.5 MHz above the video carrier. Frequency modulation is used for the audio signal, and amplitude modulation with a vestigial lower sideband is used for the video signal.

The total spectrum width of a broadcast television channel thus is the sum of the vestigial video sideband (1.25 MHz), the audio carrier (4.5 MHz), the audio upper sideband (0.05 MHz), and the guard band (0.20 MHz). This total bandwidth is 6.0 MHz.

MODULATED VIDEO CARRIER WAVEFORM

The composite video information is transmitted as amplitude modulation of a radio frequency (RF) carrier. The modulating signal is the composite video waveform in which black has the maximum amplitude and white has the minimum amplitude, as shown in Figure 4-2. This is called negative transmission, since logically white would be plotted in the positive direction.

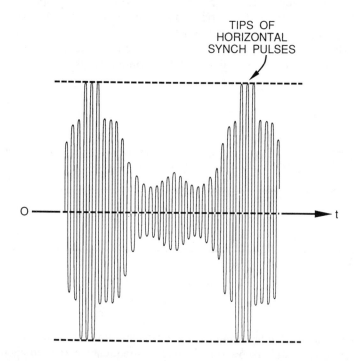

Fig. 4-2 Depiction of the video waveform used to amplitude-modulate a high-frequency radio carrier. The carrier is a considerably higher frequency than drawn here. With amplitude modulation, the negative portion of the waveform is a mirror image of the postive portion. The information is carried in the peaks of the carrier which is called the envelope.

One advantage of negative transmission is that pulses of noise in the amplitude-modulated carrier waveform cause peaks in the direction of black. This makes the noise less noticeable. Another advantage is that less transmitter power is required. This is because typical pictures are mostly white for which the amplitude of the modulated carrier is small. Lastly, the maximum amplitude of the modulated carrier corresponds to the tips of the sync pulses, and thus can be used as a reference level in the receiver that is independent of the luminance signal. This reference level can be used to control automatic gain control circuits.

CHANNEL ALLOCATION

Television broadcasting operates at radio frequencies which mostly propagate along straight lines of sight. This means that the television radio waves usually do not propagate more than about 30 miles before a physical obstacle, such as a mountain or the curvature of the earth, impedes their path. Since two major metropolitan areas can easily be closer than 60 miles, the television stations operating in each area must use different frequency bands to minimize interference.

The FCC specifies and assigns the different frequency bands to the various broadcasters serving an area. These assignments are made to minimize interference between stations operating in close adjacent areas.

A television station transmits two separate signals: one for the composite video signal and another for the audio signal. The video signal is transmitted on an amplitude-modulated carrier situated 1.25 MHz above the lowest frequency in the assigned band. The audio signal is transmitted on a frequency-modulated carrier situated 4.5 MHz above the video carrier.

Television signals are broadcast in both the very-high frequency (VHF) band and the ultra-high frequency (UHF) band. The VHF band is subdivided into two regions: the low-band VHF channels 2 to 6 extending from 54 MHz to 88 MHz and the high-band VHF channels 7 to 13 extending from 174 MHz to 216 MHz. Each television channel occupies 6 MHz in the radio spectrum.

As is shown in Figure 4-3, channel 2 in the VHF band extends from 54 to 60 MHz; channel 3 from 60 to 66 MHz; and channel 4 from 66 to 72 MHz. The spectrum space from 72 to 76 MHz is used for other purposes, such as air navigation. Channel 5 extends from 76 to 82 MHz, and channel 6 extends from 82 to 88 MHz. This completes the low-band VHF channels. The space from 88 to 108 MHz is used for FM radio transmission. The space from 108 to 174 MHz is used for other purposes, such

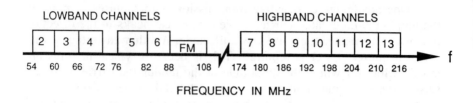

FREQUENCY IN MHz

Fig. 4-3 The frequency bands allocated to VHF television channels.

as mobile and emergency radio communications. The high-band VHF television channels 7 to 13 begin at 174 MHz and extend evenly every 6 MHz until the upper bound of 216 MHz is reached.

The UHF television channels numbered 14 to 83 extend from 470 MHz to 890 MHz. Each channel is located every 6 MHz within this spectrum range.

PROPAGATION PROBLEMS

The very-high frequency radio waves used for television propagate in straight lines from the transmitting antenna, thereby limiting transmission mostly to line-of-sight distances. These high-frequency radio waves are very directional and are easily reflected off surfaces, such as the exterior walls of large buildings. The result of such reflections is that in addition to the signal that travels directly to the receiving antenna, one or more reflected signals also travel to the receiving antenna *via* a single path or multiple paths, as shown in Figure 4-4. The reflected signals travel a slightly greater distance in reaching the receiving antenna and hence arrive somewhat later than the signal that travels the direct path. These reflected signals create distinct overlapping images, called "ghosts," on the picture.

Radio waves travel very nearly at the speed of light, 186,000 miles per second. If a reflected path is 0.5 mile longer than the direct path, the reflected signal will arrive about 3 μs later. Since a horizontal scan line is 63.5 μs long, the reflected signal will be visible as a ghost shifted to the right by about 1/20 of the horizontal width of the screen.

A reflected signal might arrive so closely after the direct-path signal that a distinct ghost would not be produced. The effect, however, could then be to blur sharp vertical edges in the picture. With color-television transmission, the result of reflections can be to create false colors in the picture by interference generated with the 3.58 MHz signal used to encode the color information.

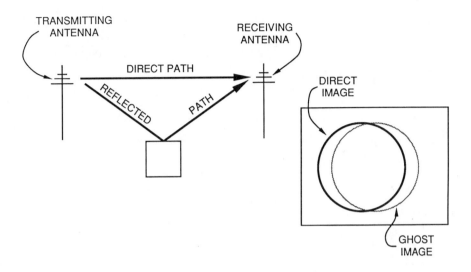

Fig. 4-4 **The very high frequency radio waves used for television can easily bounce or reflect off a surface, thereby creating two or more multiple paths from the transmitting antenna to the receiver. These multiple paths result in ghosts in the picture.**

The use of digital processing has been proposed for television receivers and is described in a later chapter. One advantage of such digital processing is that ghosts could be eliminated from the picture by calculating signals that are exact replicas of the reflected signals and by then subtracting them from the received signal.

The video signal is sent as an amplitude-modulated radio-frequency carrier. The information is contained in the variations of the peaks of the carrier, called the *envelope*. Noise affects mostly the amplitude variation of the modulated carrier, thereby strongly disturbing the information-bearing envelope. Upon demodulation, the noise appears in the video signal. Thus, the video signal is highly prone to the efects of noise, such as electric motors and automobile ignition systems. The audio signal is sent as a frequency-modulated carrier, and, since the information is encoded as frequency changes, additive noise has much less effect on the demodulated audio signal.

5. Colorimetry

LIGHT SPECTRUM

If white light is passed through a prism, the different wavelengths in the light are all bent at slightly different angles. The result is a linear display of all the different colors that compose the visible light spectrum.

The visible light spectrum ranges from violet and blue through green and yellow to orange and red. The visible light spectrum can be displayed along an axis corresponding to the dominant wavelength of light for each specific color, as shown in Figure 5-1. The wavelengths for visible light range from 380 to 770 nm. The abbreviation nm refers to a nanometer, or 10^{-9} meter (m). Wavelength is also frequently given in μ, the abbreviation for a millimicrometer, also equal to 10^{-9} meter.

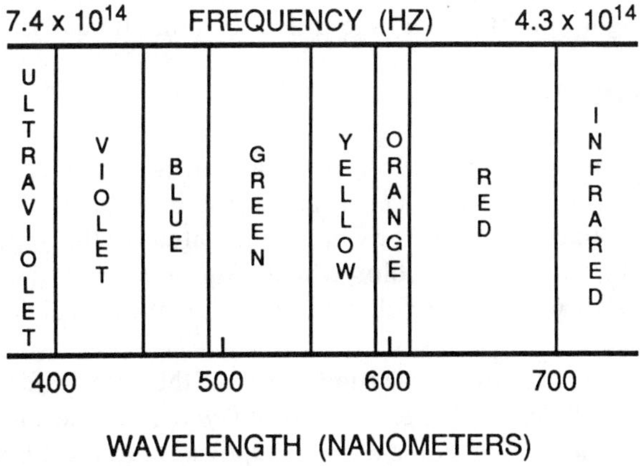

WAVELENGTH (NANOMETERS)

Fig. 5-1 The visible color spectrum.

Wavelength equals the velocity of a wave divided by its frequency. The velocity of light in a vacuum is constant at very nearly 3×10^8 meters per second (m/s). There thus is a one-to-one correspondence between wavelength and frequency. The frequency of light corresponding to wavelengths at the violet end of the spectrum is about 7×10^{14} Hz and to wavelengths at the red end of the spectrum is about 4×10^{14} Hz.

COLOR THEORY

The principles of operation of color television are firmly based on the psychophysical properties of human color vision. The laws describing human color vision were first stated by Herman Grassman of Germany in 1854. The properties of human color vision allow certain compromises to be made without which compatible color television would not have been possible. Hence, it is useful to review the theory of color vision to understand better the principles of color television.

The *additive tristimulus theory* of color vision states that any color can be matched by an additive combination of different amounts of three *primary colors*. Primary colors are unique and cannot be created from any other primaries. The three additive primary colors are red, green, and blue. White and gray are created by an additive combination of equal amounts of the three primary colors. The technical term for grays is *achromatic*.

The additive primary colors are not the same as the subtractive primary colors. The subtractive primaries apply when paint pigments are used, since the pigments absorb, or subtract, color from the white light that illuminates them. In color television, the phosphors used in the picture tube emit light, and thus additive primaries must be used.

COLOR WHEEL

The various colors that can be created from the three primaries can be arrayed around a *color wheel,* as shown in Figure 5-2. The specific "color," technically called the *hue,* is indicated by the specific point on the circumference on the color wheel. The center of the color wheel corresponds to white. The color wheel is sometimes called the *Munsell hue circle.*

A color located on the circumference of the color wheel is a "vivid" color, technically called a *highly saturated color.* Moving along the radius toward the center of the color wheel is equivalent to the addition of white to the highly saturated color on the circumference. The highly saturated color thus becomes less saturated and more pastel as more white is added to it.

The color diagonally opposite a primary color is called a *complementary color.* A complementary color produces the color white when added to a primary color. The complementary color to the primary color red is the greenish-blue color known as cyan. Yellow is the complementary color to blue, and the bluish-red color known as magenta is the complementary color to green.

COLOR ATTRIBUTES

Thus far, two attributes of the visual sensation of color have been described. The intrinsic nature of color, such as red, yellow, or cyan, is

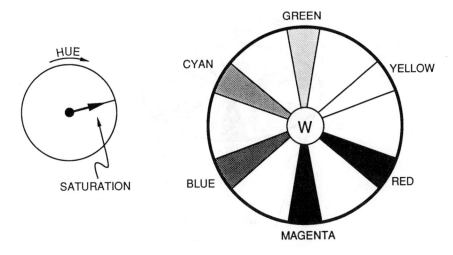

Fig. 5-2 The color wheel. The distance around the circumference indicates the hue, and the distance along the radius indicates the saturation. The center represents white.

called *hue*. The measure of color intensity, pastel *versus* vivid, is called *saturation*. The third, and last, attribute of visual sensation is the amount of light energy, called *brightness*.

Corresponding to each of the three attributes of visual sensation are three physical characteristics of color. The *dominant wavelength* is the physical characteristic that corresponds to hue, and *purity* corresponds to saturation. Hue and saturation taken together define color, and their combination is called *chrominance*. The physical characteristic corresponding to brightness is the *luminous flux*. Luminous flux is also called luminance.

The three attributes of visual sensation can be combined in a three-dimensional space to create a cone, as shown in Figure 5-3. The apex of the cone corresponds to the absence of any sensation, or black. The base of the cone corresponds to the brightest color sensations. A two-dimensional slice through the cone produces a color wheel corresponding to a particular brightness sensation.

In 1931, the Commission Internationale de l'Eclairage (CIE) adopted international standards for color definition and measurement. The additive primary colors selected as standards by the CIE are red at a wavelength of 700 nm, green at 546.1 nm, and blue at 435.8 nm. The three primary colors specified by the FCC for use in color television are also red, green, and blue. Based on the practical nonmonochromatic phosphors used in

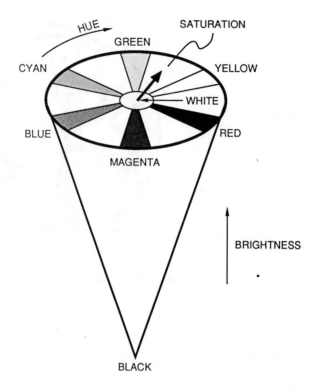

Fig. 5-3 Three-dimensional color cone.

television picture tubes, however, the wavelengths are specified as 610, 534, and 472 nm, respectively.

COLOR SPATIAL RESOLUTION

For most visual images, most of the fine detail is conveyed by changes in brightness, or luminance. For small objects, the human eye is not very good at identifying color, and most of the response is to changes in brightness. Thus, for very small areas in a scene, the human eye responds mostly to monochromatic changes in brightness. For intermediate-sized areas, the eye responds mostly to the colors cyan and orange. For large areas, the eye responds to all colors. All this is similar to color comic strips in which the color information is used only for large areas, and black lines are used for the finer details.

CIE CHROMATICITY DIAGRAM

The relative energies $\bar{r}$, $\bar{g}$, and $\bar{b}$ of the three CIE primary colors of red, green, and blue to match a pure or highly saturated color at a particular monochromatic wavelength can be plotted as a function of wavelength, as shown in Figure 5-4. These plots of the CIE tristimulus values were obtained by experiments using human subjects to match unknown colors. The negative values in the curves are areas in which the primary color is added to the unknown color to achieve a perfect match.

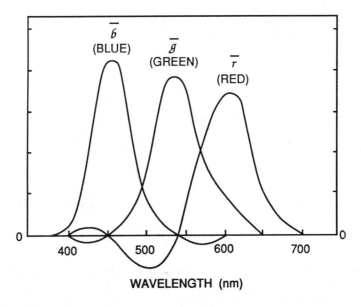

Fig. 5-4 The number of watts of each CIE primary color needed to match one watt of any color. [*Source:* Schure (1975), Plate 6.]

Clearly, negative values of a primary color are not practical in most real color-matching cases. Hence, the relative amounts of the three primary colors defined by the curves $\bar{r}$, $\bar{g}$, and $\bar{b}$ are transformed to a new set of positive tristimulus value curves $\bar{x}$, $\bar{y}$, and $\bar{z}$ for three artificial, or nonphysical, primaries, shown in Figure 5-5. The tristimulus value $\bar{y}$ is the standard luminosity function for white light and represents the spectral response of the human eye. The areas under each tristimulus curve are

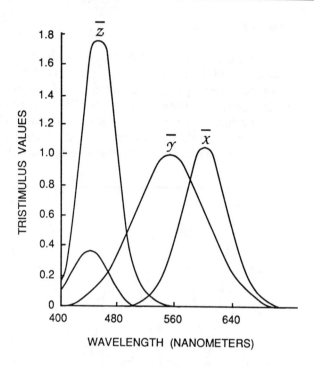

Fig. 5-5 Tristimulus values for the three artificial tristimulus primary colors.

all equal, since for white light equal amounts of each primary color are required.

The tristimulus values can be plotted in a three-dimensional space. In this way, all color hues and all saturations can be represented in terms of the amounts of the three primaries needed to create them. The problem, however, is that three-dimensional representations of data are difficult to depict for any real usefulness. A two-dimensional representation thus is required. This representation is the CIE chromaticity diagram which is created as follows.

The specific tristimulus values to match a pure or highly saturated color at a particular monochromatic wavelength are obtained from the tristimulus curves $\bar{x}$, $\bar{y}$, and $\bar{z}$ as their values at the particular wavelength. The tristimulus values are symbolized as X, Y, and Z. The tristimulus values X, Y, and Z *are each divided by their sum $X + Y + Z$ to produce* normalized chromaticity coordinates x, y, and z in a three-dimensional space:

$$x = X/(X + Y + Z),$$

$$y = Y/(X + Y + Z),$$

and

$$z = Z/(X + Y + Z).$$

The *locus* of all x, y, and z in the three-dimensional space is a tilted plane representing all colors of equal brightness.

Clearly, the sum of the chromaticity coordinates equals one,

$$X + Y + Z = 1.$$

Hence one coordinate is redundant and can be expressed in terms of the other two, as follows:

$$Z = 1 - X - Y.$$

Thus, a three-dimensional space defined by x, y, and z has been reduced to a two-dimensional plane defined by only x and y. A plot of the two-dimensional chromaticity coordinates x and y for all possible colors is called the CIE chromaticity diagram, as shown in Figure 5-6.

The CIE chromaticity diagram represents hue (dominant wavelength) and saturation (purity). The tristimulus value Y is needed to specify the luminance. The dominant wavelength of the vivid completely saturated spectral colors is indicated on a horseshoe curve. The line closing this curve represents the nonspectral hues of a range of purples. White is located at a point in the center of the CIE chromaticity diagram. The saturation of any color is measured as the distance from the white point to the location of the saturated color on the horseshoe curve. The CIE primary colors of red, green, and blue are indicated on the diagram along with the NTSC primary colors. All colors enclosed within a triangle connecting the *loci* of the three primary colors can be matched by an appropriate combination of the three primaries.

The tristimulus values X, Y, and Z of a pure or highly saturated color were found as the intersection of the tristimulus value curves $\bar{x}$, $\bar{y}$, and $\bar{z}$ with the monochromatic wavelength. The procedure is more complicated if the color sample has a continuous spectrum. Then the spectral energy distribution of the sample is multiplied by each of the tristimulus curves, and the area under each of the three resulting curves is the tristimulus value X, Y, and Z.

The next chapter applies the preceding color theory to color television, truly a marvel of analog technology.

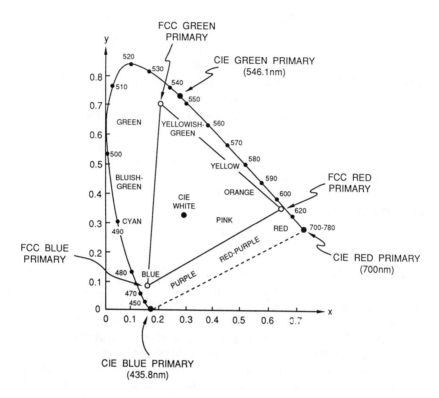

Fig. 5-6 The CIE chromaticity diagram.

References

Billmeyer, Fred W., Jr., and Max Saltzman, *Principles of Color Technology (Second Edition),* Wiley (New York), 1981.

"Color Perception," Hazeltine Corporation, Report No. 7125, April 4, 1952.

Neal, C. Bailey, "Television Colorimetry for Receiver Engineers," *IEEE Trans. Broadcast Television Receivers,* Vol. BTR-19, August 1973, pp. 149–162.

Pritchard, D. H., "US Color Television Fundamentals — A Review," *IEEE Trans. Consum. Electron.,* Vol. CE-23, November 1977, pp. 467–478.

Schure, Alexander, *Basic Television — Volume 6,* Hayden (Rochelle Park, NJ), 1975, pp. 6-114 to 6-133.

The Science of Color, Optical Society of America (Washington, DC), 1963.

6. *Principles of Color Television*

INTRODUCTION AND OVERVIEW

Color television is truly a marvel of technology. The additional information required for the specification of the color of the television picture was ingeniously inserted along with the black-and-white information, thereby maintaining compatibility with monochrome television. A new picture tube was devised to display the color picture on the home television receiver. New color cameras were developed for the studio. All this has been accomplished at prices affordable by nearly all consumers.

The first broadcast color television system, approved in October 1950 by the FCC, was a *field-sequential system* invented by CBS engineers. The picture was scanned by a monochrome television camera. A rotating disk with transparent color sectors in red, green, and blue was placed in front of the camera lens. The disk rotated at a rate such that one colored sector was in front of the lens for exactly one field. The result was a sequence of fields corresponding to the primary color components of red, green, and blue in the image. At the television receiver, another color-filter disk rotated in front of the picture tube in exact synchrony with the three fields. The color wheel was divided into three transparent color filters in red, green, and blue, corresponding to the information in each of the
three color fields.

The field rate for the field-sequential color television system was 144 fields per second resulting in 24 complete color frames per second. The number of lines per frames was reduced from the 525 lines used in monochrome television to 405 lines, and the horizontal scanning rate was 29,160 lines per second. Clearly, this system was not compatible with monochrome television (*see* Reitan, p. 53).

The television receiver for the field-sequential system was quite bulky because of the rotating color filter, and exact mechanical synchronization had to be maintained between the rotating filter and the fields presented on the picture tube. Also, and perhaps most important, the system was not compatible with existing monochrome black-and-white television receivers.

The problem of maintaining compatibility was solved by RCA's engineers, and the color television system which they and the National Television System Committee (NTSC) submitted was adopted as the national standard by the FCC in December 1953. The CBS field-sequential system was abandoned.

The NTSC color television system ingeniously encoded the red, green, and blue signals in such a way that compatibility with existing monochrome television receivers was obtained. This was accomplished by dividing the color television signal into two components. One component contained the monochromatic luminance information, and the other component contained the color or chrominance information. The chrominance signal was ignored by monochrome receivers, and only the luminance signal was displayed on the picture tube. Color television receivers decoded the additional chrominance signal to create the primary-color signals needed to create a color picture on an all-electronic color cathode ray tube. That this compatibility could be obtained within the existing bandwidth allocated for each television signal was truly amazing and even today would represent the pinnacle of analog consumer-electronics technology.

For small objects, the human eye loses its ability to distinguish color, and the only response is to changes in luminance. This means that considerably less bandwidth is needed to encode the chrominance information in a picture. Nevertheless, additional bandwidth is needed for the chrominance information. Actually, the luminance and the chrominance signals are able to share the same bandwidth because both signals exhibit a regular harmonic structure. By offsetting the chrominance spectrum exactly the right amount in the frequency spectrum, its harmonics are made to fall exactly between the harmonics of the luminance signal. This technique is called *frequency interleaving*.

The first step in encoding the color television signal is to form *color difference signals*. The color difference signals are formed by subtracting the luminance information Y from each primary color signal, R, G, and B. The color difference signals are typically small, since most of the information in an image is in the luminance information. One color difference signal actually is redundant, and hence only two color difference signals are required. The NTSC color television system uses the $(R - Y)$ and $(B - Y)$ color difference signals. These two signals are combined to create two new signals, called the I and the Q signals (to denote *in-phase* and *quadrature*), which contain all the chrominance information. The chrominance information is finally used to modulate a high-frequency color subcarrier at about 3.5 MHz simultaneously in amplitude and in phase. The amplitude modulation of the color subcarrier encodes the saturation information, and the phase modulation encodes the hue information.

The color television receiver decodes the chrominance information contained in the amplitude and phase modulation of the color subcarrier.

The chrominance information is then combined with the luminance information to obtain the three primary color signals which are finally applied to the three electron guns of the color picture tube. The electron beams emitted by each gun strike different color phosphors on the screen finally creating a color image based on the three additive primary colors of red, green, and blue.

More details about the principles of color television are explained in the following sections of this chapter.

SIGNAL SPECIFICS

The color television camera at the studio creates three signals R, G, and B corresponding to the amounts of red, green, and blue in the image being scanned. If each of these signals were given the full bandwidth used by a single monochrome signal, then a total bandwidth three times the monochrome bandwidth would be required. Clearly, compatibility with conventional broadcast monochrome television would be lost with such a scheme. The color television standard proposed by the National Television System Committee (NTSC) in 1953 avoids this problem and indeed maintains compatibility with monochrome television.

The three primary color signals are combined to create a luminance signal Y that contains the monochrome brightness information according to the following formula:

$$Y = 0.30R + 0.59G + 0.11B.$$

The different amounts of each primary signal are needed because of different sensitivities of the human eye to different primary colors. The human eye is most sensitive to green and then to red and to blue, in this order. For white light, the three primary signals would all be equal and would equal the luminance signal, or $R = G = B = Y$.

When there is no color information in the scene being scanned, i.e., when the scene is gray, it is desirable that the color signals being broadcast along with the luminance signal should vanish. Since most scenes consist primarily of gray luminance, there is little color information, and hence the color signals should likewise be small. This can be accomplished by using color difference signals which are formed by subtracting the luminance signal Y from each of the primary color signals: $(R - Y)$, $(G - Y)$, and $(B - Y)$. Since many images consist mostly of noncolored gray information, the color difference signals will mostly vanish, and very little color information will need to be transmitted.

There are now four equations but only three unknowns. This means that one of the quantities is superfluous. With some algebraic manipulation, it is possible to show that any one of the color difference signals can be expressed in terms of the other two. In particular,

$$(G - Y) = -0.51(R - Y) - 0.19(B - Y).$$

Hence, all the information in the scanned image can be specified by three signals: Y, $(R - Y)$, and $(B - Y)$. The first signal is called the luminance signal. The last two signals contain all the chrominance information and are called the chrominance signals. The color difference signal $(R - Y)$ varies from red to bluish-green, and the color difference signal $(B - Y)$ varies from blue to greenish-yellow.

The primary signals can be reconstructed from the two color difference signals and the Y signal according to the following equations:

$$R = (R - Y) + Y,$$
$$G = -0.51(R - Y) - 0.19(B - Y) + Y,$$

and

$$B = (B - Y) + Y.$$

Two new signals are formed from the color difference signals. These two new signals are formed as linear combinations of the two color difference signals in such a way that one of the new signals contains information about colors ranging on the color wheel from orange to cyan, while the other new signal contains information about colors ranging from magenta to yellow-green. The first new signal is called the I signal and is formed as follows:

$$I = 0.74(R - Y) - 0.27(B - Y).$$

Since the human eye is responsive to spatial frequencies less than about 1.5 MHz for the colors encoded by the I signal, the I signal is band-limited to 1.5 MHz.

The second new signal is called the Q signal and is formed as follows:

$$Q = 0.48(R - Y) + 0.41(B - Y).$$

The human eye is responsive to spatial frequencies less than 0.5 MHz for the colors encoded by the Q signal, and therefore the Q signal is band-limited to 0.5 MHz. For spatial frequencies greater than 1.5 MHz, the human eye does not resolve color very well, and hence all the gray information needed for the picture is contained in the Y signal.

The I and the Q signals can also be constructed directly from the R, G, and B signals according to the following equations:

$$I = 0.60R - 0.28G - 0.32B,$$

and

$$Q = 0.21R - 0.52G + 0.31B.$$

The R, G, and B signals need to be reconstructed at the telvision receiver from the received I, Q, and Y signals. This is accomplished according to the following equations:

$$R = 0.96I + 0.63Q + Y,$$
$$G = -0.28I - 0.64Q + Y,$$

and

$$B = -1.11I + 1.72Q + Y.$$

The I and Q signals are used to modulate two color subcarriers at the same frequency but 90° out of phase with respect to each other. Another term for a phase difference of 90° is "quadrature," hence the use of the "Q" for the Q signal. The I signal is the "in-phase" signal, hence the use of the "I" to describe it. The luminance and the modulated subcarrier chrominance signals are added to produce the final signal that is used to modulate the radio-frequency (RF) carrier for broadcast over the air. Mathematically, this can be expressed as:

$$M = Y + [Q \sin(Ft + 33°) + I \cos(Ft + 33°)],$$

where F is the frequency of the color subcarrier and t is the time. The term in brackets is called the chrominance signal, C, and thus $M = Y + C$.

Some trigonometry is useful to help understand the preceding form of subcarrier modulation. The frequency of the subcarrier is symbolized as F in the preceding equation. The maximum amplitude Q of the cosine term is the x-axis projection of a radius of length $|C|$ at some angle ϕ with respect to the x-axis. Similarly, the maximum amplitude I of the sine term is the y-axis projection of the same radius. By Pythagorus's theorem, the magnitude of the hypotenuse $|C|$ of the right triangle formed by the x-axis and the y-axis projections is the square root of the sum of the squares of the projections, or, as shown in Figure 6-1,

$$|C| = \sqrt{Q^2 + I^2}.$$

The angle ϕ of the radius is the angle having its tangent given by Q/I. A quantity that varies in both magnitude and phase is called a vector.

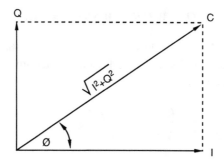

Fig. 6-1 Vector addition of *I* and *Q* signals to form a single *C* signal that varies in amplitude and phase.

Thus, another way of visualizing the modulation of the subcarrier is that the maximum amplitude, or magnitude, $|C|$ of the subcarrier varies with the sum of the squares of *I* and *Q*, while simultaneously the phase of the subcarrier varies with the ratio of *Q* to *I*. This is equivalent to simultaneous amplitude modulation and phase modulation of the subcarrier. Stated mathematically,

$$M = Y + [|C| \sin(Ft + 33° + \phi)],$$

where *F* is the frequency of the color subcarrier, *t* represents time,

$$|C| = \sqrt{I^2 + Q^2},$$

and

$$\phi = \tan^{-1}(Q/I).$$

Since both *Q* and *I* vanish for the white and grays, the maximum amplitude or magnitude of the subcarrier indicates the saturation of the color information. The phase of the subcarrier specifies the hue of the color information.

For white light and grays, the color difference signals $(R - Y)$ and $(B - Y)$ are equal to zero, or vanish. Since the I and Q signals depend only on these two color difference signals, they too vanish. Hence, the magnitude $|C|$ of the chrominance signal vanishes, and there is no color subcarrier present. Thus, this type of quadrature modulation results in a suppressed carrier.

The 33° phase shift of both the I and Q signals affects their relative phase compared to the reference color burst. The reason for this is to simplify the color demodulation circuits in the television receiver.

WAVEFORMS

The chrominance information is encoded as the simultaneous amplitude and phase modulation of a 3.579545 MHz subcarrier that is added to the luminance signal. The waveform for a horizontal scan line thus is a high-frequency sine wave varying in average value, maximum amplitude, and phase. The average value represents the luminance signal, Y. The varying maximum amplitude of the sine wave represents the saturation of the color information. The varying phase of the sine wave represents the hue of the color information.

Phase is difficult to determine for a signal without a ready reference. The phase reference is obtained from a short burst of about eight full cycles of the color subcarrier that is placed on the back porch of each horizontal blanking signal, as shown in Figure 6-2. This burst of color subcarrier with a known reference phase is called the *color burst*.

As will be explained later, the horizontal scanning frequency for color television had to be changed from the 15,750 Hz used for monochrome television to 15,734.264 Hz. Thus a horizontal scan line for color television is 1/15,734.264 seconds, or 63.5 μs, long. Since the back porch contains the color burst, the back porch for color television is lengthened from its value for monochrome television, and this necessitates a change in the width of the blanking and sync pulses. The blanking and sync pulses for color television are 10.5 and 4.8 μs wide respectively to allow more time for the back porch. The front porch is 1.3 μs, and the back porch is 4.4 μs.

The frequency of the color subcarrier, 3.579545 MHz, was chosen to be an odd harmonic of half the horizontal scanning frequency of 15,734.264 Hz. Thus, in the time period of one full horizontal scan line there are exactly (3,579,545)/(15,734.264) or 227.5 cycles of the color subcarrier.

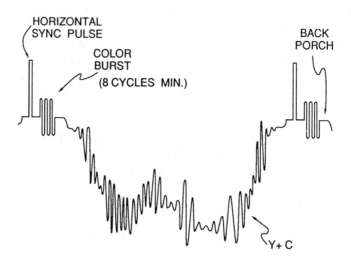

Fig. 6-2 Composite video waveform for a single horizontal scan line of the color video signal. The color, or chrominance information, is carried by a high-frequency subcarrier sine wave that is varied in amplitude and phase with respect to the reference color burst. The amplitude variation represents saturation, and the phase variation represents hue. The average value is the luminance information.

This means that the following scan line begins exactly one-half cycle out of phase with respect to the preceding scan line, as shown in Figure 6-3. Because of the closeness of the scan lines and also the persistence of the phosphors used in the picture tube, the high-frequency visual effects of the color subcarrier on the luminance information cancel when averaged across two adjacent scan lines. Thus, compatibility with monochrome receivers is maintained.

COLOR VECTOR DIAGRAM

The phase relationships between the various signals used to encode color information can be shown in a two-dimensional color vector diagram, as shown in Figure 6-4. The x-axis is the $(B - Y)$ signal, and the y-axis is the $(R - Y)$ signal. The Q signal is along an axis inclined 33° to the x-axis, and the I signal is along a radial line perpendicular to the Q axis. The color reference has a phase that is 180° out of phase with respect to the $(B - Y)$ signal and thus is shown along the negative $(B - Y)$ axis.

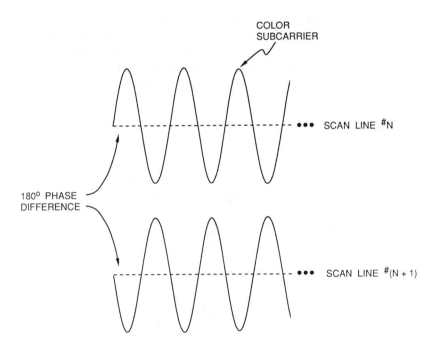

Fig. 6-3 The frequency of the color subcarrier was chosen to ensure a 180° phase difference between adjacent scan lines, thereby canceling the visibility of the color signal on the television screen.

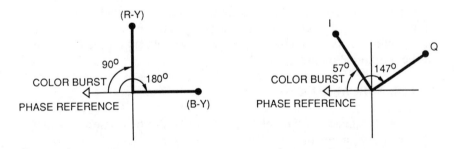

Fig. 6-4 Vector diagrams showing the phase relationships among the various signals used in color television. The color burst is the phase reference of 180° with respect to the $(B - Y)$ signal.

The vector positions corresponding to the three primary colors red, green, and blue can also be shown on the vector diagram, as in Figure

6-5. At the television studio, a vector oscilloscope is used to display the color signal to be sure that the signal meets FCC standards. A standard test pattern with maximally saturated primary colors is used, and the amplitude and phase of the signal corresponding to these colors should then fall within specified areas in the display on the vector oscilloscope.

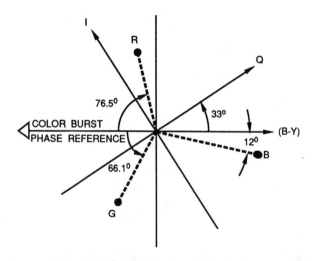

Fig. 6-5 Vector diagram showing the approximate positions of the red, green, and blue signals in relation to the color burst and other color signals.

Figure 6-6 shows the approximate colors corresponding to a series of vectors spaced every 30° with respect to the phase reference. The phase reference is the phase of the color burst placed on the back porch of the horizontal blanking signal.

COLOR SUBCARRIER

The frequency of the color subcarrier had to be chosen to satisfy a number of requirements.

The color subcarrier had to be within the luminance band, yet should not be visible as a high-frequency spatial signal on monochrome receivers. This was accomplished by ensuring that the color subcarrier was 180° out of phase on adjacent horizontal scan lines so that it would thereby cancel on monochrome receivers. For this to occur, the frequency of the color subcarrier was chosen to be an odd harmonic of half the horizontal scanning frequency.

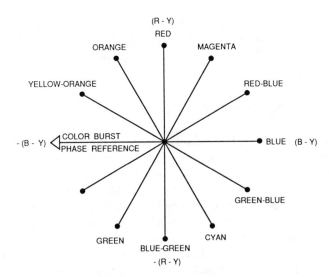

Fig. 6-6 Vector diagram of the color space represented by changes of 30° in the phase of the chrominance signal.

The bandwidth of the Q signal is limited to 0.5 MHz above and 0.5 MHz below the subcarrier frequency. The I signal, however, has a bandwidth of 1.5 MHz below the subcarrier frequency and a vestigial bandwidth 0.5 MHz above the subcarrier frequency. Since the color information was to be placed at the higher frequency portion of the video band, the color subcarrier would thus need to be about 0.6 MHz below the upper video frequency of 4.2 MHz. Thus, the color subcarrier would need to be in the vicinity of 3.6 MHz.

The audio carrier at 4.5 MHz could not be changed without causing major changes in the circuitry of existing monochrome television receivers. It was, however, important that the color subcarrier avoid any objectionable frequency beats with the audio carrier. This was accomplished by choosing a horizontal line frequency, the 286th harmonic of which was 4.5 MHz. This horizontal line frequency was 15,734.264 Hz. Although this was a slight change from the horizontal line frequency of 15,750 Hz used previously for monochrome transmission, the horizontal sweep circuits used in monochrome television receivers easily had enough latitude to lock onto the horizontal sweep pulses at the slightly lower frequency. Since the number of horizontal scan lines per frame was kept at 525, the vertical scanning frequency was changed slightly to 59.94 Hz, which likewise was well within the latitude of the vertical sweep circuits in monochrome television receivers.

The requirements on the color subcarrier are that it must be an odd harmonic of half the horizontal frequency and also that it should be in the vicinity of 3.6 MHz. The chosen frequency of 3.579545 MHz met these requirements.

FREQUENCY INTERLEAVING

An essential aspect of color television is the way in which the color information coexists with the monochrome information in the frequency spectrum. This coexistence is accomplished through frequency interleaving of the harmonic structure of the color information between the harmonic structure of the monochrome information, as explained below.

Assume that some signal s(t) is perfectly periodic and a single period repeats itself at a rate F_H Hz. The frequency spectrum of such a periodic signal will consist of lines at integer, or harmonic, multiples of F_H.

A luminance video signal is highly repetitive, since each scan line is very similar to each preceding line. The frequency spectrum of such a highly repetitive signal exhibits the harmonic structure typical of periodic signals. Hence, the spectrum consists of bursts of energy at harmonic multiples of the horizontal scanning frequency of roughly 15,750 Hz.

Of course, each scan line is not the same for a real television signal, even if the image itself is stationary. The scan lines for a stationary image change as the image is scanned from top to bottom and then scanned again and again at the field and frame rates. The frequency spectrum for such a "still" image has components at harmonic multiples of the scanning frequency F_H, but each of these components has smaller lines above and below, in essence, small sidebands. The spacing between these smaller lines is the field and frame rate, or 60 and 30 Hz. If the image changes with respect to time, the major harmonic lines and their smaller brethren develop sidebands and become diffuse.

Thus, the frequency spectrum of a television signal consists of bursts of energy at harmonic multiples of the horizontal scanning rate. Each burst of energy itself exhibits a symmetric harmonic structure at multiples of the frame and field rates of 30 and 60 Hz, as shown in Figure 6-7. Each burst of energy decays to zero after a dozen or so harmonic multiples of the frame rate which is well before the beginning of the next burst of energy centered around the next harmonic of the horizontal scanning frequency. What this means is that there is a fair amount of empty space in the frequency spectrum between the bursts of energy centered around the harmonics of the horizontal scanning frequency. The frequency spectrum looks like the periodically spaced teeth of a comb.

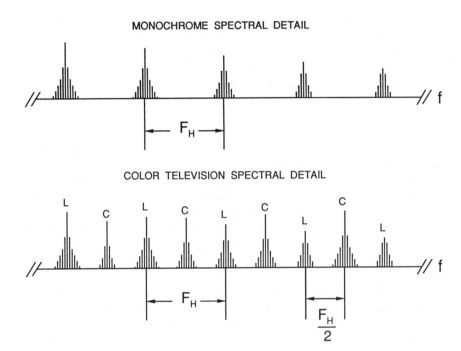

MONOCHROME SPECTRAL DETAIL

COLOR TELEVISION SPECTRAL DETAIL

Fig. 6-7 The spectrum of a monochrome television signal consists of a number of bursts of energy located at harmonic multiples of the horizontal scanning frequency, F_H. The space between these bursts is empty. In color television, the spectrum of the color information is fitted in these empty spaces.

The chrominance signal likewise is fairly repetitive at the horizontal scanning frequency, and thus it, too, has a frequency spectrum that exhibits a harmonic structure very much like the frequency spectrum of the luminance signal. Thus by offsetting the frequency spectrum of the chrominance signal by half the horizontal scanning frequency, its spectral bursts of energy fall exactly in between the spectral bursts of the luminance signal. In this way, the two signals share the same spectral space with no interference between them, as shown in Figure 6-7. This technique of interleaving two signals in the frequency spectrum is called frequency interleaving.

The harmonic nature of the frequency spectrum of a television signal was discovered in the early 1930s by researchers at Bell Labs. Mathematical analyses and frequency analyses of actual images confirmed their discovery that the frequency spectrum of a scanned image consisted of alternating bands of high energy at multiples of the scanning frequency and regions of little energy. The researchers realized back then that the idle space in the spectrum could be used for independent signals. It was not until the idle space was used for the chrominance signal, however, that their discovery was put to real use.

FREQUENCY SPECTRUM

The chrominance information is centered around the color subcarrier of 3.579545 MHz. The horizontal scanning frequency of 15,734.264 was chosen so that the color subcarrier is an odd harmonic of half the horizontal scanning frequency, actually the 455th harmonic. In this way, the bursts of energy of the chrominance signal fall precisely in between the bursts of energy of the luminance signal. Since most images have most of their energy in the lower frequencies, the energy in the luminance signal is decreasing as the energy in the chrominance signal is increasing in the frequency spectrum.

The chrominance information is encoded as simultaneous amplitude and phase modulation of the color subcarrier frequency of 3.579545 MHz. This modulation is accomplished by quadrature amplitude modulation of a sine wave and a cosine wave by the information contained in the Q signal and in the I signal. The modulation of the color subcarrier by the Q signal generates an upper and a lower sideband centered at the color subcarrier frequency. The Q signal is band-limited to 0.5 MHz, and hence the upper and lower sidebands each extend 0.5 MHz above and below the color subcarrier. The modulation of the color subcarrier by the I signal likewise creates an upper and a lower sideband, except that the I signal is band-limited to 1.5 MHz. Hence its lower sideband extends 1.5 MHz below the color subcarrier. The upper sideband is filtered so that only a vestige remains extending 0.5 MHz above the color subcarrier.

The frequency spectrum of a color television signal thus consists of three components: the luminance signal, the I signal, and the Q signal. These three components occupy a spectrum space that extends 4.2 MHz above the video RF carrier with vestigial transmission extending 1.25 MHz below the video RF carrier. The audio signal is transmitted as a frequency modulated audio RF carrier that is 4.5 MHz above the video RF carrier, which is identical to the audio scheme used in monochrome television transmission. The spectrum of a color television channel is shown in Figure 6-8.

The luminance signal in color television transmission occupies the same spectrum space as the monochrome luminance signal in monochrome television, from 1.25 MHz below the video RF carrier to 4.2 MHz above the video RF carrier. The I and the Q signals occupy a spectrum

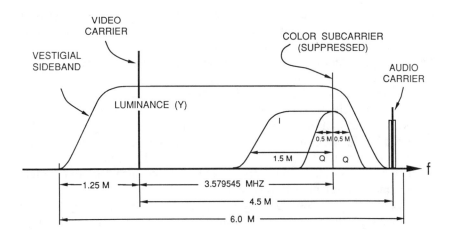

Fig. 6-8 The spectrum of a color television channel. The Q signal is band-limited to frequencies below 0.5 MHz, and the I signal is band-limited to frequencies below 1.5 MHz. The upper sideband of the I signal is limited to 0.5 MHz.

space centered around the color subcarrier, which is 3.579545 MHz above the video RF carrier.

A problem with the NTSC color television system is that the luminance signal and the chrominance signal share the same overall space in the frequency spectrum. Although frequency interleaving theoretically allows the two signals to be kept distinct in the spectrum, the method is not without its problems. One problem is that special filters, called comb filters, are needed to separate totally the harmonic structure of one signal from the other. Another problem is that for certain moving images the harmonic clusters become wide enough that frequency interleaving no longer separates the luminance and chrominance signals. Then one signal interferes with the other, and the two signals cannot be completely separated from each other.

The interference between the luminance and chrominance signals creates two possible problems: cross-color interference and dot crawl. With cross-color interference, the high-frequency harmonic detail in the luminance signal contaminates the chrominance information and creates color beat patterns. These beat patterns are visible on the sharp edges of objects in a scene and on repetitive patterns like the herringbone pattern of a tweed jacket. With dot crawl, the chrominance subcarrier signal contaminates the luminance information and becomes visible as a dot pattern that crawls up the screen. The use of comb filters at the receiver enables almost complete separation of the luminance information from the chrominance information, thereby eliminating the contamination of one signal by the other because of inadequate filters. Appropriate prefiltering at the camera and transmitter can reduce the interference effects encountered with certain moving images.

Another series of problems encountered with television arises from the use of interlaced scanning which can result in visible flicker under certain viewing and image conditions. As one example, large-area flicker can occur if the television image is viewed peripherally. Certain repetitive patterns can create moiré images on the screen. Interline flicker can be seen along the horizontal borders of objects, since such horizontal edges are scanned and displayed only 30 times per second.

Large-screen television receivers are popular. If such large screens are viewed too closely, individual scan lines can be seen, and flicker might also become noticeable.

Research is under way to eliminate or reduce these and other problems encountered with conventional NTSC television. The motivation for much of this research activity is high-definition television (HDTV) which is discussed in a later chapter.

References

Hirsch, C. J., W. F. Bailey, and B. D. Loughlin, "Principles of NTSC Compatible Color Television," *Electronics,* February 1952, pp. 88–95.

Howard W. Sams Editorial Staff, *Color-TV Training Manual,* Howard W. Sams (Indianapolis), 1980, pp. 23–46 and 99–121.

Mertz, Pierre, and Frank Gray, "A Theory of Scanning and Its Relation to the Characteristics of the Transmitted Signal in Telephotography and Television," *Bell System Tech. J.,* Vol. 13, 1934, pp. 464–515.

Pritchard, D. H., "US Color Television Fundamentals — A Review," *IEEE Trans. Consum. Electron.,* Vol. CE-23, November 1977, pp. 467–478.

Wentworth, John W., "Basics of Color Video," *Broadcast Engineering,* March 1979, pp. 82–93.

7. Picture and Camera Tubes

PICTURE TUBES

The single most expensive component in a television receiver, and in many respects the most important component, is the picture tube. The received television image is displayed on the picture tube for viewing. Nearly all displays in television receivers utilize cathode ray tubes (CRTs) to display the picture, with the exception of very small screen television receivers that use liquid crystal displays (LCDs). The cathode ray tube has indeed withstood the test of time, and even in this current high-technology age has no equal. The CRT produces a bright, clear picture with virtually no discernible flicker. The CRT is robust and very affordable when manufactured in the vast quantities required for television receivers.

MONOCHROME CATHODE RAY TUBES

The basic operational principles of the cathode ray tube are fairly straightforward. An electron beam is focused and accelerated until it impacts on the phosphor coating on the inside of the faceplate of the tube. The energy of the electrons bombarding the phosphor raise the energy level of the electrons in the atoms of the phosphor crystals. When the excited electrons return to their initial energy state they emit quanta of light energy in the process. The light emitted during excitation by the electron beam is called fluorescence, and the light emitted after excitation is called phosphorescence.

The brightness of the excited phosphor depends on such factors as the actual phosphor used to coat the screen, the voltage used to accelerate the electron beam, the beam current, and the duration of the excitation. The spectral characteristics of the light emitted depend on the specific phosphor and can be made to vary over the the entire visible band through choice of appropriate phosphors. The phosphor used in a monochrome receiver emits white light, while the phosphors used in color picture tubes emit red, green, or blue light. The persistence of the light emitted by the phosphor should not be too long, or smearing of the picture will result. Typical phosphors used in monochrome CRTs have persistences of about 5 ms.

The faceplate of most CRTs used in television receivers is aluminized. This process results in a fine coating of aluminum over the phosphor coating on the inside of the faceplate of the tube. The electrons from the electron beam easily pass through the aluminum coating and bombard the phosphor as usual. The light from the rear of the phosphor, however, is reflected back through the phosphor by the reflective aluminum coating. This nearly doubles the light output from the tube.

The faceplate is frequently made from "gray" glass. The ambient light in the room passes once through the gray glass and then again when it is reflected off the white phosphor. The light emitted from the excited phosphor, however, passes only once through the gray glass. If the glass faceplate passes only 1/2 of the light, then the reflected room light will be reduced by two passes through the faceplate by a factor of $(1/2) \times (1/2)$ or 1/4. The light emitted from the phosphor, however, passes only once through the faceplate and is reduced by a factor of 1/2. The net effect is to improve the contrast of the television image in brightly lit rooms by a factor of two for this example.

The electron beam is produced by an assembly of electrodes in the neck of the tube, as shown in Figure 7-1. This assembly of electrodes is called the electron gun. The electrodes themselves are shaped like little cups. The electrons are boiled from the surface of a heated electrode, called the *cathode*, through a process called *thermionic emission*. The cathode is heated by a filament of wire, called the *heater,* with a heater voltage of 6.3 volts. Most CRTs use quick-heating cathodes so that a picture is produced quite rapidly when the receiver is initially turned on.

The *control grid* has a small hole in its center through which the electrons can pass. The voltage on the control grid is negative with respect to the cathode. By varying this negative voltage, it is possible to control the number of electrons passing through the hole. In some televison receiver circuits, the control grid is used to control the overall brightness of the picture. In other receiver circuits, the video signal itself is applied to the control grid to modulate the electron beam with the video information.

The *screen grid* is positive with respect to the cathode and attracts electrons. It serves the purpose of accelerating the electrons in the beam. The focus grid is shaped in such a way that the electrostatic field along its axis focuses the electron beam when it strikes the phosphor on the face of the tube.

The inside surface of the sides of the tube are coated with a black graphite material called *aquadag*. This coating forms the anode to attract the electron beam toward the face of the tube. The anode is at a very high positive voltage relative to the cathode. This voltage, called the *ultor voltage,* typically is from 10 to 20 kV for monochrome CRTs. It is about 25 kV for color CRTs.

The inside of the tube is evacuated of air to form a vacuum. The glass portion of the tube is called the *envelope*. As was previously explained, the faceplate of many tubes is tinted neutral gray to reduce the

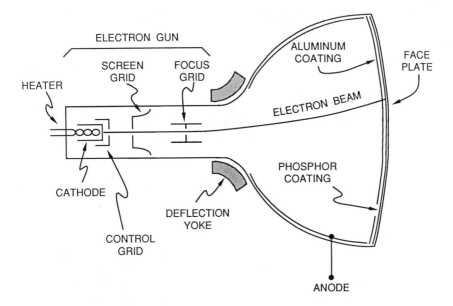

Fig. 7-1 Cutaway side view of monochrome cathode ray tube. The electrons are "boiled" from the surface of the cathode and are attracted to the very high positive voltage surrounding the faceplate. The rear of the faceplate is coated with phosphor which emits light when struck by the electron beam. The electrical current in coils of wire placed around the neck of the tube creates varying magnetic fields that interact with the moving beam of electrons to cause vertical and horizontal deflection of the beam.

effect of ambient light on the picture. Since the tube is a vacuum and the envelope is glass, there is a danger that the tube might implode if physically damaged, shattering glass on the viewer. One solution is use of a safety panel bonded to the faceplate.

The electron beam is deflected vertically and horizontally by coils of wire wound around the neck of the tube. Electrical current flowing in the coils creates a magnetic field that interacts with the moving electrons in the beam to cause the beam to deflect. This method of deflection is called *electromagnetic deflection*. The coils of wire are called the *vertical* and *horizontal deflection yokes*.

The size of picture tubes is given as the corner-to-corner diagonal length of the image on the faceplate in inches. To decrease the depth of television cabinets, the maximum deflection of the electron beam is increased. This deflection is measured as the total top-to-bottom deflection of the beam and ranges from 90° to 110° for most CRTs. The higher angles require more power for the deflection yokes, which has become a consideration from the perspective of energy conservation.

COLOR PICTURE TUBES

The basic principle of generating a color picture is the use of a cathode ray tube with three electron beams corresponding to the three primary colors of red, green, and blue. The phosphors used to coat the rear of the faceplate are organized into many small arrays of red, green, and blue light–emitting phosphors. Each electron beam strikes only its corresponding phosphor in each small array, thereby causing the appropriate amount of primary light to be emitted. The three beams sweep together across the face of the screen.

The three electron beams are created in such a way that they each strike the screen from a slightly different angle. A metal mask consisting of many small holes or slits is placed slightly behind the coating of phosphors on the rear of the faceplate. Thus, only those portions of the electron beams that pass through the small apertures actually strike the phosphors. This mask is called the *shadow mask*. An example of a shadow mask with two sources of light is shown in Figure 7-2.

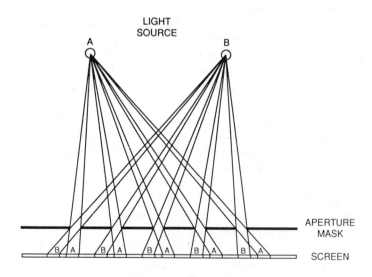

Fig. 7-2 Principle of operation of the shadow mask color tube. In this example, two light sources pass through slits in an aperture mask, thereby causing alternating bands of light on the screen. In an actual color display tube, three electron beams are used as the sources.

The three electron beams are produced by different physical arrangements of three electron guns. The *electron gun* is the name given to the total assembly of the cathode, control grid, accelerating grid, and any focusing grids. In one arrangement, shown in Figure 7-3, the three guns are arranged in a *delta pattern*. The delta pattern of electron guns is matched by a delta dot pattern of red, green, and blue phosphors on the screen. In another arrangement, shown in Figure 7-4, the three electron guns are arranged along a horizontal line. This arrangement is called *in-line guns* and is matched by vertical stripes of the three phosphors on the screen.

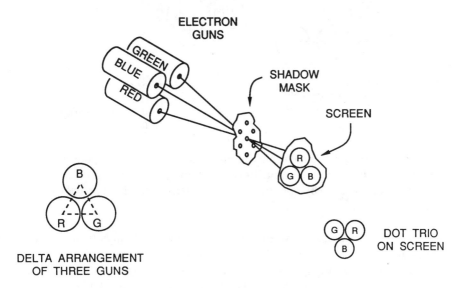

Fig. 7-3 Depiction of three electron guns arranged in a delta pattern for a color picture tube. The result after passing through the shadow mask is a series of red-green-blue dot trios on the screen.

There is a third arrangement, shown in Figure 7-5, that is used with some color picture tubes. A single electron gun with three cathodes in a line is used in this arrangement. The three cathodes all share a common control grid, screen or accelerating grid, and focus grid. The three electron beams are focused by the focus grid and then caused to converge at the shadow mask through electrostatic convergence by convergence plates. The shadow mask consists of vertical slits through which the three beams pass. The three primary-color phosphors are arranged as groupings of three stripes on the screen.

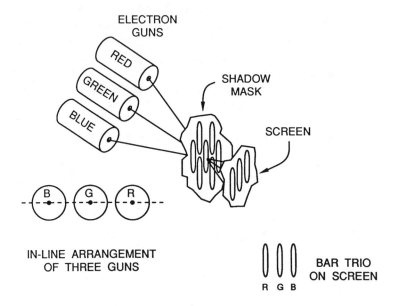

ELECTRON
GUNS

RED

GREEN

BLUE

SHADOW
MASK

SCREEN

B   G   R

IN-LINE ARRANGEMENT
OF THREE GUNS

BAR TRIO
ON SCREEN

R   G   B

Fig. 7-4 Depiction of a color picture tube using three electron guns arranged in line along the same horizontal line. The shadow mask consists of a number of vertical slits, thereby creating a series of red-green-blue bar trios on the screen.

The three electron beams are deflected electromagnetically by deflection coils wound around the neck of the tube. Unfortunately, the three beams converge in a spherical plane, as shown in Figure 7-6, and hence are not converged at the periphery of the shadow mask. The solution is to apply correction signals to special coils placed around the neck of the tube. This type of correction is called *dynamic convergence* of the beams. The dynamic convergence of picture tubes with in-line guns is much simpler than that of tubes with the delta arrangement of the three guns.

In addition to dynamic convergence of the beams, fine tuning of the beams is needed to assure that each beam strikes its corresponding phosphor at all points on the screen. This type of *static convergence* is performed by adjustments of small magnets located near the neck of the tube. Convergence of a color picture tube is a complicated affair to be performed only by professionals.

The project to develop the shadow mask color picture tube was initiated by engineers at the RCA Laboratories in 1949, and the first working protoype was available the following year. Large-scale production began in 1954.

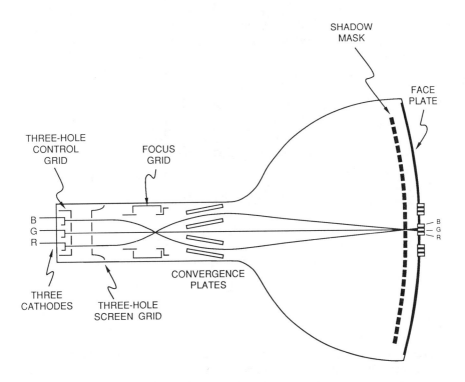

Fig. 7-5 Depiction of a color picture tube using a single set of electrodes for accelerating, focusing, and converging the three electron beams. The three beams are created by three cathodes. This type of single-gun color picture tube is used in television receivers manufactured by the Sony Corporation.

The Trinitron color picture tube developed by the Sony Corporation was the first tube to use in-line electron guns. This tube had a large electron lens to keep the electron spots small, and the in-line guns simplified convergence. The Trinitron tube used vertical slits with no horizontal supporting members. This improved vertical resolution, but required a heavier, deeper tube. In 1972, RCA announced its color picture tube with precision in-line electron guns. Vertical slots with horizontal supporting webs were used so that the shadow mask could be curved spherically, thereby simplifying convergence over the whole face of the tube and resulting in a lighter, shallower tube.

In 1969, both RCA and Zenith introduced color picture tubes with a "black matrix." The black-matrix tubes used smaller phosphor dots with a black area between them. The result was improved contrast of the

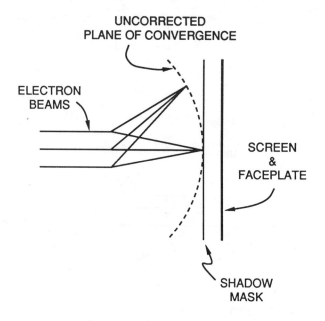

Fig. 7-6 Side view of the uncorrected path of the three electron beams. The three beams converge along a circular arc that does not correspond to the placement of the shadow mask and screen. Correction of this convergence problem is achieved through the use of static and dynamic magnets and magnetic circuits placed near the neck of the color tube.

image, since ambient room light fell on the black matrix and was not reflected back at the viewer.

GAMMA CORRECTION

As is shown in Figure 7-7, the light output from a cathode ray tube is not linear with the applied signal voltage. Expressed mathematically, the light output L equals the signal voltage V applied to the tube raised to a power γ (the Greek character, gamma), or

$$L = V^{\gamma}.$$

Gamma has a value of 2.2 for color picture tubes and a value of approximately 2.5 for black-and-white tubes. The *gamma effect* compresses blacks and stretches whites.

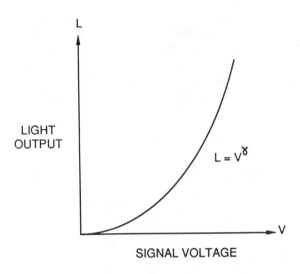

Fig. 7-7 The light output from a picture tube is a nonlinear function of the signal voltage. The relationship is exponential. The exponent is gamma, thereby leading to the name *gamma correction* to correct the problem.

A problem is that a strict proportionality must be maintained between the amounts of the red, green, and blue signals, or the colors displayed on the picture tube will be distorted. Hence, the R, G, and B signals transmitted from the television studio must be modified to correct for the gamma effect in the picture tube. The gamma-corrected signals are sometimes symbolized with primes, as R', G', and B'. The equations for Y, the color difference signals, I, and Q would all use the gamma-corrected primary-color signals.

A result of the gamma-correction is that the gamma-corrected luminance signal is no longer a true luminance signal for monochrome receivers. It is, however, a reasonably good approximation. The only distortion is that highly saturated reds and blues are reproduced in a black-and-white receiver as grays that are too dark.

CAMERA TUBES

The earliest method for converting an image into an electrical signal was the electromechanical Nipkow disk, previously described in Chapter 2. This method was impractical for modern television, for which an all-electronic technique for scanning an image was needed. The solution was the electronic *image–orthicon tube*, first introduced commercially by RCA in 1941, in which an electron beam was swept across a photoconductive surface. The image–orthicon tube was replaced in 1951 by the *vidicon tube* which is more compact and less complex. Most broadcast television cameras use *Plumbicon* or *Saticon tubes* which have the same construction as the vidicon tube but have targets that operate on a different principle.

The vidicon tube generates a varying electrical current in proportion to the varying amounts of light in an image. This current then produces the varying voltage needed for the video signal.

The image is focused on the target area at the rear of the faceplate of the tube using conventional optics. As shown in Figure 7-8, the rear of the faceplate is coated with a transparent layer of electrically conducting material, usually tin-oxide. A layer of photoresistive material, such as antimony sulfide, is placed over the transparent conductor. The light image passes through the transparent conductor and strikes the photoresistive material, causing it to be electrically charged in proportion to the amount of light striking it. An electron beam scans the charge image on the rear of the photoresistive material, thereby creating an electrical current that is proportional to the charge.

The electron beam is created by a heated cathode and is accelerated by an accelerator grid. The beam is focused electrostatically by a focus grid within the tube and also electromagnetically by exterior coils around the neck of the tube. The beam is deflected vertically and horizontally by electromagnetic deflection using exterior deflection coils around the neck of the tube. A low dc voltage of 30 to 50 volts (V) is placed between the target and the cathode.

The photoresistive material is, in effect, a variable resistance that has a higher resistance of about 20 million ohms ($M\Omega$) in the black portions of the image and a lower resistance of about 2 $M\Omega$ in the white portions of the image. The electron beam flows through these variable resistance portions of the photoresistive image, thereby creating a variation in the electrical current. This varying electrical current flows through a fixed load resistor, creating a varying voltage which is then amplified to produce the video signal.

The signal that results when there is no light on the vidicon tube is very small. This gives an excellent reference black level for the video signal. The amount of current flowing when the tube is dark is called the *dark current*.

In most television studios, the horizontal and vertical deflection of all the cameras is synchronized with a single, common source. In some cameras, the vertical and horizontal sweep signals are supplied externally

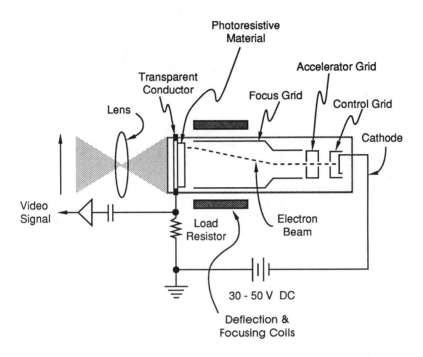

Fig. 7-8 The image vidicon camera tube is used to generate a video signal. The scene is focused by a lens onto a photoresistive material. The resistance of the material decreases with the amount of light falling on it. An electron beam sweeps the photoresistive image resulting in an electrical current proportional to the amount of light energy stored in the photoresistive material.

from the common source. In more recent cameras, synchronization signals are supplied externally and the cameras generate their own sweep signals internally. Since the sweep generators of all the cameras are locked together in synchrony, the technique is called "gen-lock."

COLOR CAMERAS

The simplest approach to creating a color camera is the use of three image tubes and three primary-color filters. A single lens focuses the image, and partially reflective mirrors direct the image through each of the three filters to the tubes, as shown in Figure 7-9. One problem with this type of color camera is that one-third of the light is lost while passing through the mirrors.

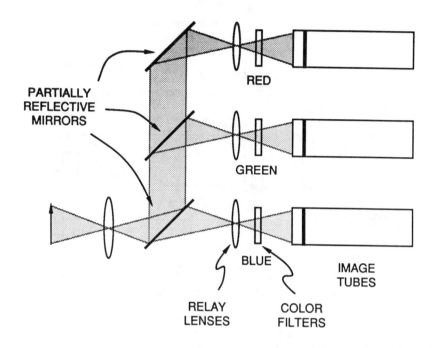

PARTIALLY
REFLECTIVE
MIRRORS

RED

GREEN

BLUE

IMAGE
TUBES

RELAY
LENSES

COLOR
FILTERS

Fig. 7-9 Three image tubes are used in a color camera. Each camera responds to a different primary color. Mirrors and color filters separate the light from the scene into the three color components for the three image tubes. Newer color cameras use prisms, rather than mirrors, to create the three images.

An improvement in the beam splitting is obtained through the use of dichroic mirrors and reflecting surfaces. A dichroic mirror reflects light of one wavelength but passes light of all other wavelengths. Thus, either dichroic mirrors or prisms can be used to separate the light into its three primary components for the three image tubes.

The output of the three tubes is gamma corrected through the use of nonlinear amplifiers. The three gammma-corrected primary signals are used as inputs to a matrix where the luminance and chrominance signals are calculated, as depicted in Figure 7-10. The two components of the chrominance signal are used to quadrature modulate the color subcarrier which is then added to the luminance signal.

Newer television cameras utilize solid-state imagers that offer such advantages as less power, smaller size, and greater reliability compared to image tubes. Solid-state imagers are described later in a chapter on new display and camera technology.

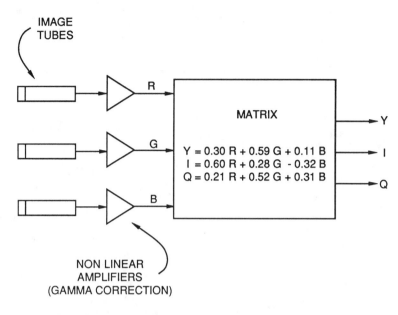

IMAGE
TUBES

R

MATRIX

Y

Y = 0.30 R + 0.59 G + 0.11 B
I = 0.60 R + 0.28 G - 0.32 B
Q = 0.21 R + 0.52 G + 0.31 B

G

I

B

Q

NON LINEAR
AMPLIFIERS
(GAMMA CORRECTION)

Fig. 7-10 The signals from the three image tubes are combined in a circuit called a matrix to create the Y, I, and Q signals used to encode the luminance and chrominance information for broadcast over the air.

References

Grob, Bernard, *Basic Television and Video Systems (Fifth Edition)*, McGraw-Hill (New York), 1984, pp. 77–102.

Herold, Edward W., "History and Development of the Color Picture Tube," *Proc. SID,* Vol. 15, 1974, pp. 141–149.

McGinty, Gerald P., *Video Cameras: Theory and Servicing,* Howard W. Sams (Indianapolis, IN), 1984, pp. 19–23.

McMann, Renville H., Jr., "Television Cameras," in *Electronics Engineers' Handbook (Second Edition),* Donald G. Fink (Editor-in-Chief), McGraw-Hill (New York), 1982, pp. 20-24 to 20-42.

Morrell, A. M., "Color Picture Tube Design Trends," *Proc. SID,* Vol. 22, 1981, pp. 3–9.

8. Receiver Circuitry

FUNCTIONAL BLOCK DIAGRAM

The circuitry of a television receiver can be organized into a small number of different functional elements or blocks, as shown in Figure 8-1. The first block of circuitry is the so-called *"front-end"* where the radio frequency signal is received and demodulated. The front-end is also called the *tuner*. The *audio block* demodulates and amplifies the sound signal. The *video block* amplifies the video signal in a monochrome receiver and also performs the color separation in a color receiver. The *sweep block* generates the signals required for the vertical and horizontal deflection of the display. The front-end, audio, and sweep circuits are mostly the same in both monochrome and color receivers.

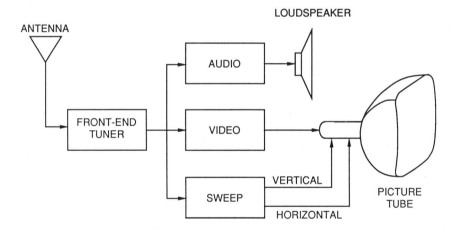

Fig. 8-1 Block diagram of major functional circuits in a television receiver.

Figure 8-2 shows a more detailed diagram of the various circuit elements in a monochrome receiver. The following sections of this chapter describe these circuits. The circuitry of a color receiver is described toward the end of the chapter.

FRONT-END CIRCUITS

The television band of signals is received by the VHF and UHF antennas which produce a voltage proportional to the received electro-magnetic field. This received voltage is very small and also contains all

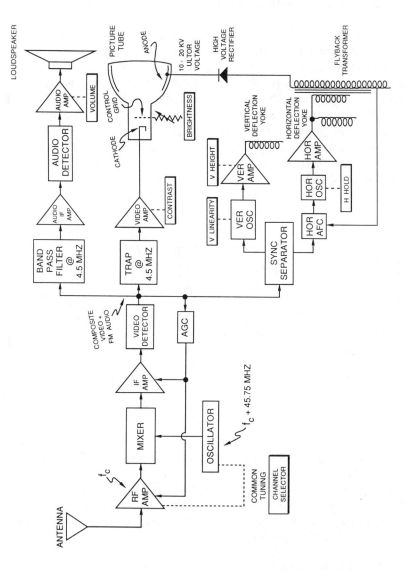

Fig. 8-2 Detailed schematic of the circuitry of a monochrome television receiver.

the channels in the TV band that has been selected, either the VHF band or the UHF band. The 6-MHz band of frequencies corresponding to the desired channel is band-pass filtered from all the other channels and is amplified by a tunable RF amplifier. The RF signal in the chosen band is then mixed with a sine wave generated by a tunable oscillator. This oscillator generates a sine wave with a frequency equal to the video RF carrier plus 45.75 MHz. The bandpass filters in the RF amplifier and the tunable oscillator are both controlled by the channel selector.

The mixer multiplies the RF signal by the sine wave. The result of this process is a new signal that contains all the sum and difference frequencies of the two input signals. Through filtering, only the difference frequencies are passed. The RF amplifier, oscillator, and mixer taken together as a single unit are called the tuner.

The RF signal contains frequencies from 1.25 MHz below the RF carrier to 4.75 MHz above the RF carrier, i.e., $f_c - 1.25$ to $f_c + 4.75$ MHz. The frequency of the tunable oscillator is $f_c + 45.75$ MHz. The band of frequencies represented in the difference band thus cancels the RF carrier frequency and range from 41 to 47 MHz. This new band of frequencies is called the intermediate frequency band, or the *IF band*. The IF band is inverted in the sense that the audio carrier has been translated to 41.25 MHz at the lower end of the IF band and the video carrier has been translated to 45.75 MHz at the upper end of the IF band.

The result of the mixing operation is that all channels, regardless of their particular RF band, have been converted, or beat down, into the same IF band from 41 to 47 MHz. The IF band is amplified by the IF amplifier. The IF amplifier includes a sharp filter to eliminate the signals from any adjacent channels.

The amplification of the RF and IF amplifiers is controlled by an *automatic gain control* (AGC) circuit, the purpose of which is to ensure that the output of the IF amplifier does not vary for different channels. Since the sync pulse amplitude remains constant, it is sensed and used by the AGC circuit to control the gain of the RF and IF amplifiers.

After passing through the mixer, the video information is encoded as amplitude modulation of a 45.75 MHz carrier. This video IF signal is demodulated by the video detector to create the baseband video signal.

AUDIO CIRCUITS

The demodulation process of the video IF signal results in a mixing of the audio and video carriers so that sum and difference frequencies are again created. This produces a frequency-modulated signal at 45.75 MHz − 41.25 MHz = 4.5 MHz that contains the audio information. This frequency-modulated 4.5 MHz signal is called the *audio IF signal*, and it is extracted from the demodulated video IF signal by a bandpass filter centered at 4.5 MHz.

The audio IF signal is amplified by the audio IF amplifier prior to recovery of the audio baseband signal by the audio detector. The audio signal then is amplified and finally applied to the loudspeaker, where it is converted from electrical energy to an acoustic signal. The volume control controls the volume of the audio output from the loudspeaker by varying the gain of the audio amplifier.

VIDEO CIRCUITS

The baseband video signal is amplified by the video amplifier to a level near 85 volts. The input to the video amplifier is preceded by a filter that removes the 4.5 MHz audio IF signal to prevent any possible interference in the picture. This type of filter is called a "trap." The gain of the video amplifier is varied by the contrast control.

The amplified video signal is applied to either the cathode or the grid of the picture tube, depending upon the configuration used by the manufacturer of the receiver. If the video signal were applied to the cathode, the polarity of the signal would be such that the downward, or negative-going, portion corresponded to increasing amounts of white. In this way, more negative-going signal at the cathode would cause more electrons to be boiled from its surface, thereby increasing the luminance of the display. The overall brightness of the picture can be varied by controlling a constant voltage at the control grid of the picture tube.

An alternative approach is to invert the polarity of the video signal and apply it to the control grid. In this way, a positive-going signal corresponds to more white, and the increasing positive signal at the grid would attract more electrons from the cathode, thereby increasing the signal striking the screen of the picture tube.

SWEEP CIRCUITS

The output from the video detector includes both the luminance signal and the sync pulses. This is the composite video signal. This signal is used as input to the sync separator which separates the horizontal and vertical sync pulses. Because the length of the vertical sync pulses is much longer than the length of the horizontal sync pulses and also because the horizontal frequency is greatly faster than the vertical frequency, this separation process is fairly simple.

The horizontal and vertical oscillators are free-running, which means that in the absence of any sync pulses they automatically produce sweep waveforms at very nearly the proper frequencies. The sync pulses fine tune these frequencies to be in exact synchrony with the sweep rates transmitted from the television studio. An advantage of the free-running feature of the oscillators is that the picture will remain reasonably stable if the sync pulses are momentarily lost for any reason.

The separated vertical sync pulses control the frequency of the vertical oscillator. The output of the vertical oscillator is a saw-toothed wave that sweeps the beam vertically across the picture tube. The linearity of the saw-toothed wave is varied by the vertical linearity control. The saw-toothed wave is amplified by the vertical amplifier, the gain of which is varied by the vertical height control. The amplified vertical saw-toothed wave is applied to the vertical deflection yoke wrapped around the neck of the picture tube. If vertical sync is lost, the picture will roll up or down the picture screen.

The horizontal sync pulses are very narrow and sharp, and, hence, are very prone to the effects of little pulses of noise. For this reason, they cannot be used directly to control the frequency of the horizontal oscillator. The vertical sync pulses are much longer and thus are less sensitive to the effects of noise so that this type of problem is not encountered. The solution is to compare a sample of the horizontal sweep waveform with the horizontal sync pulses to correct for any noise effects that might cause a change in the horizontal sweep rate. This comparison is performed in the *horizontal automatic frequency control circuit* which creates horizontal sync pulses that are locked and stabilized in frequency and phase with the input pulses.

The stabilized horizontal sync pulses control the frequency of the horizontal oscillator that generates the horizontal saw-toothed wave. This wave is amplified by the horizontal amplifier and then applied to the horizontal deflection yoke. The horizontal hold control varies the free-running frequency of the horizontal oscillator. If horizontal sync is lost, the picture will tear horizontally across the picture screen.

The output from the horizontal amplifier is also applied to a step-up transformer that increases the voltage of the saw-toothed wave to a level of from 10,000 to 20,000 volts. This high voltage then is rectified by a diode to create the dc needed by the accelerator of the picture tube. This voltage is called the *ultor voltage*. The high-voltage transformer is called the *"flyback" transformer*. The voltage in a color television receiver must be more precisely controlled than in a monochrome receiver. For this reason, a voltage regulator is used to stabilize the output of the high-voltage rectifier.

COLOR RECEIVERS

Much of the circuitry in a color television receiver is quite similar to that in a monochrome receiver. The sync circuits are virtually identical, except that a high-voltage regulator is added at the high-voltage rectifier to increase the stability of the ultor voltage as required by color picture tubes. A diagram of the major circuits in a color television receiver is shown in Figure 8-3.

In color television receivers, the audio signal must be extracted from the video IF signal before the video detector. If not, the audio carrier of 4.5 MHz would beat with the color subcarrier of 3.58 MHz, thereby creating a beat frequency of $4.5 - 3.58 = 0.92$ MHz, which would cause interference in the final picture. The IF signal containing both the video and audio information is applied to a simple diode detector which creates the audio IF signal centered around 4.5 MHz. This signal is extracted from the video signal by a bandpass filter centered around 4.5 MHz. The audio IF signal then is demodulated to give the audio baseband signal which is amplified and used as input to the loudspeaker.

The output of the video detector passes through a bandstop filter, called a "trap," centered at 41.25 MHz to prevent the audio information from entering the video detector. The output from the video detector passes to the sync and sweep circuits to generate the vertical and horizontal deflection signals. The video signal also passes to the chroma circuitry where the color information is extracted. This circuitry will be described in more detail shortly.

The video signal includes the sum of the luminance signal Y and the chrominance signal C. The chrominance information is filtered out, leaving only the luminance signal which is then amplified to give Y. The luminance filter frequently is a low-pass filter that passes only frequencies below 3.2 MHz. Although this filters out the chrominance information, it also reduces the detail in the luminance signal which extends fully to 4.2 MHz.

An alternative approach is the use of a *"comb" filter* which extends fully to 4.2 MHz. The comb filter passes only bursts of energy centered about the harmonics of the horizontal frequency and has a frequency response that looks like the teeth of a comb, hence the name "comb" filter. Comb filters are more costly than simpler lowpass filters and hence are often used only in expensive receivers.

A comb filter consists of a delay element and an adder. As is shown in the circuit of Figure 8-4, the incoming signal is split in two. One portion

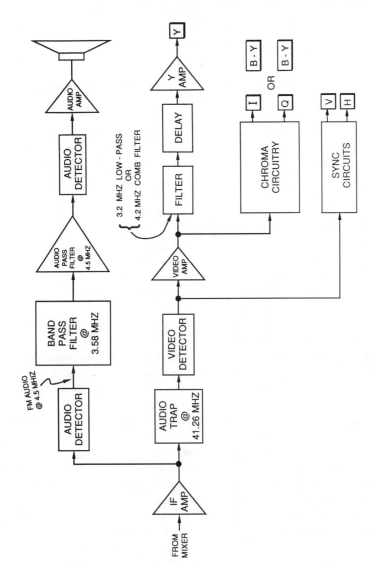

Fig. 8-3 Diagram of the major functional circuits in a color television receiver. A major change compared to monochrome circuitry is the addition of the color, or chroma, decoding circuits. Filters are also added to separate the audio, luminance, and chrominance signals.

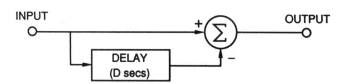

Fig. 8-4 Circuit diagram of a comb filter. The input signal is delayed and subtracted from itself.

passes through a delay of time D while the other portion goes directly to the adder. The output of the delay also goes to the adder where it is subtracted from the direct signal. At frequencies equal to $1/D$, the delayed, inverted signal will exactly cancel the direct signal. This cancellation will also occur at all harmonics of $1/D$.

For a comb filter used in a color television receiver, the delay D is exactly the length of a horizontal scan line, or $1/15,734 = 63.5$ μs. This delay is frequently performed by using a small piece of piezoelectric material, such as lithium niobate. The signal enters one end, where it causes the material to bend mechanically, thereby creating a *surface acoustic wave* (SAW) that travels to the other end. The time to traverse the distance from end to end creates the delay.

The bandwidth of the chroma circuitry is less than the bandwidth of the luminance circuitry, which results in a time delay for signals to pass through the chroma circuits. Thus, a delay must be introduced in the luminance circuits to match the chroma delay so that the demodulated color signals are in precise phase synchrony with the luminance signal. The delay is about 1 μs.

The chroma circuitry for a television receiver that demodulates the I and Q signals is shown in Figure 8-5. The chrominance information is separated from the luminance information by a filter. In some receivers, this filter is a 1-MHz bandpass filter centered at the color subcarrier of 3.58 MHz. The problem with this simple bandpass filter is that the lower sideband of the I signal extends a full 1 MHz beyond the 0.5-MHz lower cut-off of the filter, and thus color resolution is lost. An improved approach is the use of a comb filter that passes only the harmonics of the color subcarrier.

The phase reference needed to demodulate the I and Q signals is contained in the color burst inserted on the back porch of the horizontal blanking signal. The color burst is filtered from the video signal by a sharply tuned bandpass filter centered at 3.58 MHz. As shown in Figure 8-6, the color burst is then amplified by an amplifier that is only turned

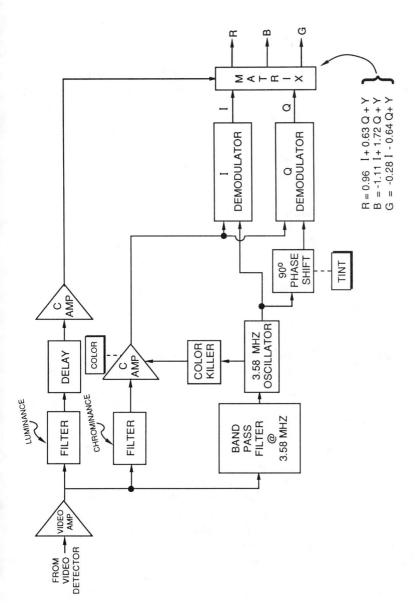

Fig. 8-5 Chroma circuitry for decoding the *I* and *Q* signals. These two signals are then combined with the *Y* signal to obtain the *R*, *G*, and *B* signals to drive the color picture tube.

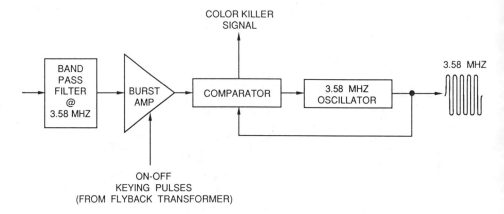

Fig. 8-6 Details of color-subcarrier oscillator circuits.

on during that portion of time that the color burst is present. A pulse from the flyback transformer is used to key on and off the color-burst amplifier. The output from the amplifier is compared with the output of a 3.58 MHz oscillator and a signal is created to lock in the frequency and phase of the oscillator precisely with the color burst. If the color burst is not present, a signal is sent to turn off the chrominance amplifier, thereby killing any spurious color in the picture.

The output of the 3.58-MHz oscillator is shifted 90° in phase and combined with the chrominance signal to perform the Q-signal demodulation. The output of the oscillator is combined directly with the chrominance signal to perform the I-signal demodulation. The 90° phase shift can be varied slightly by the tint control to change the hue of the picture.

The I, Q, and Y signals are combined linearly in a matrix circuit that performs the operations of the three equations to give the R, G, and B signals. These three signals then are applied to either the three cathodes or the three control grids of the color picture tube.

An alternative approach for the chroma circuitry, shown in Figure 8-7, is demodulation to obtain the color difference signals $(R - Y)$, $(G - Y)$, and $(B - Y)$. These difference signals could then be combined with the luminance signal Y to obtain the color signals R, G, and B which would be applied to the cathodes or the control grids of the color picture tube. Another approach is to apply the inverse of the luminance signal $(-Y)$ to the cathodes and the color difference signals to the three control grids. The subtractions between each color difference signal and the luminance signal are performed by the interactions between the cathode and the control grid, thereby resulting in electron-beam currents that are proportional to the color signals only.

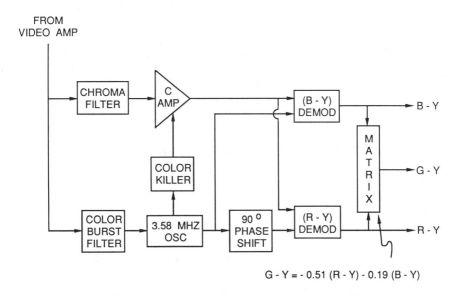

FROM
VIDEO AMP

$$G - Y = -0.51 (R - Y) - 0.19 (B - Y)$$

Fig. 8-7 Chroma circuitry for decoding the color difference signals.

COLOR DEMODULATION

The chrominance signal C contains all the information about all the various color signals depending upon the phase relationship with the color burst. The color burst is the reference phase. The chrominance signal has been filtered so that it is slowly varying in time with respect to the much faster varying color subcarrier. Sampling the chrominance signal at the appropriate delay with respect to the color burst will thus produce any desired color signal.

Figure 8-8 shows this process. Sampling the chrominance signal 90° after the reference phase gives $(R - Y)$, and sampling 180° after the reference phase gives $(B - Y)$. Sampling the chrominance signal 57° after the reference phase gives I, and sampling 147° after the reference phase gives Q. The sample values are smoothed to give continuous waveforms.

Another way of understanding this process of demodulation comes from the quadrature modulation process in which

$$C = Q \sin(Ft + 33°) + I \cos(Ft + 33°)$$

where F is the frequency of the color subcarrier. When time t is such that the quantity $(Ft + 33°)$ equals an integer multiple of 0° or 360°, the sine

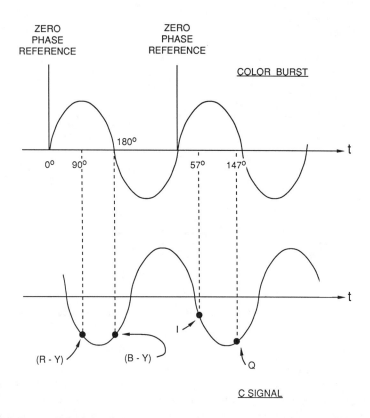

Fig. 8-8 Demodulation of the color signals through sampling of the chrominance signal at the appropriate times relative to the color burst. Sampling
every 90° and 180° relative to the color burst gives the *(R − Y)* and *(B − Y)*
color difference signals, respectively. Sampling every 57° and 147° relative
to the color burst gives the *I* and *Q* signals, respectively.

term will equal 0, since sin(0°) = 0. Then, however, the cosine term will
equal *I*, since cos(0°) = 1. Thus, sampling *C* at these instants in time gives
sample values of the *I* signal. Likewise, sampling 90° later will give sample
values of the *Q* signal, since then sin(90°) = 1 and cos(90°) = 0.

DIGITAL TELEVISION

Television receivers utilize analog technology to choose a specific
channel, to demodulate the video and audio signals, to create the video
sweep signals, and to decode the color information. In fact, all the circuitry
in a television receiver is analog. Analog integrated circuits, however,

are costly and lack the processing capabilities that digital circuits offer. For these reasons, and because of the possibilities for innovative features, there is interest in using digital circuits in television receivers.

Broadcast VHF and UHF television channels cover frequencies from 54 MHz to nearly 900 MHz, although each channel is only 6 MHz wide. These frequencies are so high that it is not presently feasible to convert the received radio-frequency (RF) signal to digital. Hence, the received RF signal is demodulated, using conventional analog circuits to give the baseband video and audio signals. These baseband signals are converted to digital and are then processed digitally in a digital television receiver.

The video signal is converted to digital by the video digital-to-analog (D/A) converter. The analog video signal is usually sampled at a rate four times the color subcarrier frequency or 14.32 million samples per second. This rate is more than twice the maximum video frequency of 4.2 MHz, thereby satisfying the Nyquist sampling theorem.

The amplitude variation of the sampled video signal is quantized into 256 levels which are then encoded using 8 bits. The use of 8 bits and 256 levels ensures an acceptable signal-to-noise ratio of nearly 50 dB. The final bit rate for the digitized video signal is 8×14.2 M $= 113.6$ million bits per second (Mb/s).

A number of new features are possible with a digital television receiver. Many of these features derive from the ability to store a whole frame of video in the receiver. A frame of video is about 1/30 s in duration. At nearly 114 Mb/s, a total of 3.8 million bits (or about 1/2 Mbyte) is required to store a whole frame. With continuing decreases in the cost of digital memory, such digital storage of a whole frame is affordable. The use of a digital frame store would enable "freeze frame" to be obtained.

A simplified block diagram of a typical digital television receiver is shown in Figure 8-9. Digital comb filters would be used to obtain the Y and C signals. These signals would then be combined digitally to obtain digital R, G, and B signals which would be converted to analog form for use by the picture tube. If some form of digital display were available, the digital R, G, and B signals would be used directly without the need of conversion to analog signals.

The sync information is also processed digitally and is used to create digital horizontal and vertical sweep signals. These digital signals are then converted to analog signals for use by the deflection yoke.

A ghost is caused by a reflected signal that arrives at the television receiver a short time after the main signal. The visual effects of a ghost can be eliminated by digital processing of the received combined signal. The ghost, or echo, is simply canceled by the digital circuitry in a manner

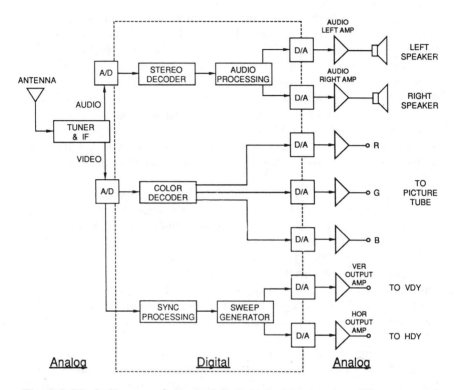

Fig. 8-9 Block diagram of a typical digital television receiver. The tuner and IF stages are analog, but all processing of the audio and video information is digital after that. The digital audio, video, and sweep signals are converted to analog signals for use by the loudspeakers, picture tube, and deflection yokes.

similar to the way audio echos are canceled on long-distance telephone circuits. Various types of digital filtering could be used to reduce the effects of received noise and other impairments.

The audio signal would also be converted to digital form and then processed to decode any stereo information. The final digital signals would be converted back to analog signals for use by the loudspeakers.

Digital circuitry enables the screen to be split so that more than one channel can be watched at the same time. It also is possible to zoom in on a feature of interest. These are only some of the features that are possible with digital television receivers. Digital circuits also improve the reliability of the television receiver. Improved separation of the Y and C signals improves the resolution of the color image.

References

Fischer, Thomas, "What Is the Impact of Digital TV?," *IEEE Trans. Consum. Electron.*, Vol. CE-28, August 1982, pp. 423–429.

Naimpally, Sam, "Digital TV," *IEEE Trans. Consum. Electron.*, Vol. CE-32, May 1986, pp. 69–76.

Smith, Kevin, "Digital TV Starts to Edge into View," *Electronics*, April 5, 1984, pp. 89–93.

9. Video Tape Recording

PRINCIPLES OF MAGNETIC TAPE RECORDING

The basic idea of recording a signal in terms of an alternating magnetic field along some magnetic material was invented in 1899 by Valdemar Poulsen of Copenhagen. Poulsen's idea truly came to fruition just before World War II, when thin tape with a fine coating of iron particles was used to record audio signals. Afterward, magnetic tape recorders were developed commercially for professional use in sound recording studios and were a key component in the commercial development of hi-fi. These professional machines used open reels of tape and were somewhat awkward to load. Magnetic tape recording for the home truly came into being with the invention by Phillips Corporation of the small audio cassette which eliminated the loading problems encountered with open-reel tape.

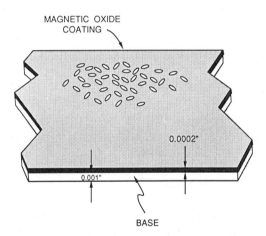

Fig. 9-1 Cross-sectional view of magnetic tape used for recording. The magnetic coating typically is about 0.0002 inch thick, and the base is about 0.001 inch thick. Open-reel audio tape is 0.25 inch wide, and audio cassette tape is 0.15 inch wide.

Magnetic tape consists of a flexible, thin base material, usually polyester, as shown in Figure 9-1. A layer of very fine magnetic particles is deposited and bonded to the base material using a binding agent such as polyurethane. The small needle-like particles are usually ferric iron oxide or chromium dioxide. Since the magnetic information is recorded longitudinally along the tape, the magnetic needles are aligned longitudinally along the tape for optimum results during the tape manufacturing process. The base is typically about 0.001 inch thick, and the magnetic coating is about 0.0002 inch thick. Open-reel audio tape is 0.25 inch wide, and audio cassette tape is 0.15 inch wide.

The changing magnetic patterns corresponding to the input signal are recorded on the tape by the *record head*. The record head consists of a near circular ring of ferrous magnetic metal with a very narrow gap that interrupts the otherwise complete ring, as shown in Figure 9-2. The gap is filled with a nonmagnetic material, such as glass or platinum. A coil of wire is wrapped around the metal. An alternating current is applied to the coil which induces a changing magnetic field in the metal ring. A changing magnetic field bridges, or fringes, around the interruption of the magnetic circuit that occurs at the gap. The magnetic tape passes by the gap, and the magnetic energy bridges this gap by traveling through the magnetic coating on the tape. A residual, or remanent, magnetism remains which is the recorded signal. The magnetization induced onto the tape is proportional to the electrical current in the coil of the recording head.

On playing back the tape, the magnetic field recorded on the tape passes by the gap. This creates a changing magnetic field that induces a voltage at the terminals of the coil. The same head can be used for recording and for reproducing the information recorded on the tape.

The relationship between the magnetizing field in the gap H and the residual, or remanent, magnetic induction B_r in the tape is highly nonlinear, as shown in Figure 9-3. This means that the magnetism in the tape would be a highly distorted replica of the signal to be recorded. There is, however, a portion of the curve which is very nearly linear. The solution thus is to operate in this linear region by adding a high-frequency tone (sine wave) to the input signal. This high-frequency wave is called the bias current.

The speed of the tape is controlled by a rotating shaft, called the capstan, which pulls the tape across the heads, as depicted in Figure 9-4. A small amount of tension is introduced at the supply reel to keep the tape in contact with the heads. The tape is wound on the take-up reel. Sometimes a separate head is used for playing the tape back, since in this way it is possible to listen, or monitor, the quality of the signal that has just been recorded immediately before on the tape. Prior to reaching the record head, the tape is erased by an *erase head*. Usually, the high-frequency bias signal is applied to the erase head to cause the erasure.

Open-reel tape recorders have tape speeds of 15, 7 1/2, and 3 3/4 inches per second (in/s). The popular cassette recorders for the home have a much slower tape speed of 1 7/8 in/s.

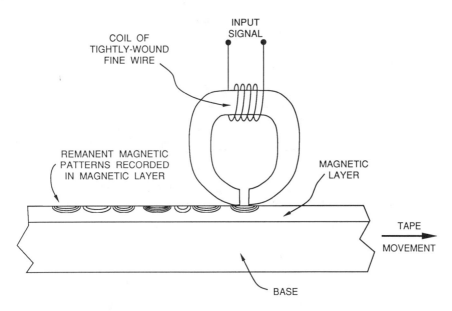

Fig. 9-2 Side view of a magnetic head used for tape recording and playback. A magnetic circuit is created in the near circular metal assembly by the signal at the coil of wire. This magnetic circuit is broken by a thin gap of non-magnetic material. The magnetic energy bridges this gap by traveling through the magnetic coating on the tape. A residual, or remanent, magnetism remains which is the recorded signal.

Magnetic tape offers the important feature of being able to erase old information to record new information. It is not without its problems, though. When the tape is wound on its reel, adjacent layers are in close contact, and the magnetic field from one layer can cause a residual magnetism in the adjacent layer. On playback, a weak signal will be heard either immediately before or after the main signal. This problem is called print through. Magnetic tape is subject to a number of physical problems, such as stretching and even breaking. Also, with passage of time the binder can deteriorate so that the oxide coating easily scrapes off the surface of the base.

Magnetic tape is a serial medium in the sense that large portions of the tape must be moved through the machine to reach a particular place on the tape. Thus, the random access of a disk medium is not possible.

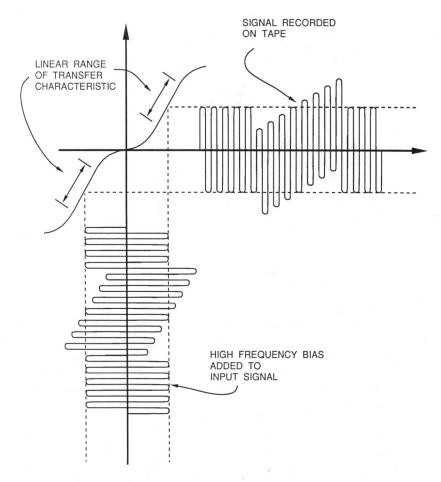

Fig. 9-3 The transfer characteristic between the magnetizing force *H* applied at the gap and the remanent magnetization *B*ᵣ in the tape is highly nonlinear. The solution is to add a high-frequency bias to the input signal to cause operation in the linear range of the characteristic.

VIDEO TAPE RECORDING PRINCIPLES

Television signals contain very high frequencies along with a dc component. These very high frequencies are impossible to record at the very slow speeds at which the tape passes by the record head as used with audio recording. The maximum frequency that can be recorded on

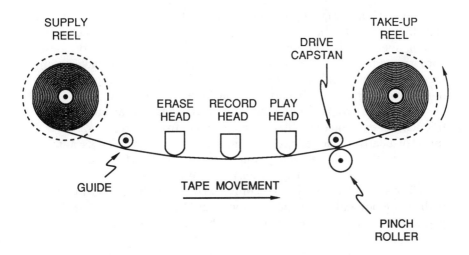

SUPPLY
REEL

DRIVE
CAPSTAN

TAKE-UP
REEL

ERASE   RECORD   PLAY
HEAD   HEAD   HEAD

GUIDE

TAPE MOVEMENT

PINCH
ROLLER

Fig. 9-4 The arrangement of magnetic heads in a tape recorder.

a tape is related to the width of the gap in the head, as is explained in the following.

The speed of the head with respect to the tape is called the *writing speed*. The frequency f of a recorded waveform is related to the writing speed v and the wavelength λ by the following relationship:

$$f = v/\lambda.$$

When the wavelength of the waveform on the tape is equal to the width of the gap, the gap bridges across the full north and south polarity of the magnetic domains, the polarities cancel, and the output is zero. Thus, the record head cannot record waveforms, the wavelengths of which are the same magnitude as the width of the gap. As a practical matter, the recorded wavelength should be no shorter than half the width of the gap.

Assume that a maximum frequency of 4 MHz is to be recorded. A practical gap width is 0.6 μm (0.6 $\times$ 10^{-6} m). The longest wavelength that could be recorded thus would be half the width of the gap, or 0.3 μm. The required speed of the tape passing by the head can be calculated from the preceding equation as the product of the frequency and the wavelength, or 4 MHz $\times$ 0.3 μm = 1.2 meters per second, which is equivalent to nearly 4 feet per second. In no way could a small cassette, or, for that matter, a practical size of open-reel tape, contain enough tape to record a reasonable amount of time at such a fast tape speed.

The solution to this problem is to move the tape at a slow transport speed, but move the heads across the surface of the tape at a much higher writing speed. This solution is used in both professional video tape recorders and home video cassette recorders. In professional machines, the record and playback heads move transversely across the tape, and in home machines the heads move diagonally across the tape.

A second problem with recording video signals is the large range of frequencies encompassed by the video signal, from 30 Hz to 4.2 MHz, a range of nearly 18 doublings in frequency, or octaves. This is much in contrast to the audio frequency range of 20 Hz to 20 kHz encountered in audio tape recorders, a range of only 10 octaves.

The strength of the signal recorded on the tape is reasonably constant with frequency until a maximum upper-limit frequency is reached. The gap width and writing speed determine this maximum frequency that can be recorded on the tape, as was explained previously.

The signal played back from the tape increases in amplitude, however, as the frequency of the recorded signal increases. This is because the speed with which the magnetic field changes determines the amplitude of the voltage at the output of the coil of wire on the playback head. A higher frequency signal recorded on the tape generates a faster changing magnetic field on playback leading to an output voltage with a greater amplitude. In fact, the amplitude of the playback signal doubles with each doubling of the frequency of the recorded constant-amplitude signal. The results of these effects on the frequency response of a typical tape recorder are shown in Figure 9-5.

A video signal encompasses a frequency range of nearly 18 octaves. Since each octave results in a doubling of the amplitude of the voltage of the playback signal, the dynamic range of the largest output voltage to the smallest output voltage would be 262,144 to 1, or 108 dB. This large a dynamic range clearly cannot be accommodated with magnetic tape recording, and the frequencies at the lower end of the band would fall below the noise level in the system. The practical range of frequencies that can be recorded and played back is 10 octaves.

The solution to this problem is to move the video signal to a higher band of frequencies through appropriate modulation of a carrier tone, or sine wave. For example, professional video recorders move the video signal into a band of frequencies from about 2 to 10 MHz through the process of frequency modulation of a carrier at roughly 6 MHz. This shifted band of frequencies encompasses a range of less than 3 octaves and thus is easily accommodated by the tape recording process.

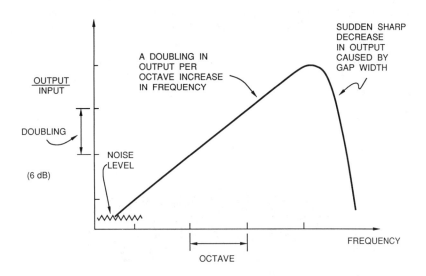

Fig. 9-5 Typical shape of the frequency response of a tape recorder. Since the output of the playback, or reproduce, head depends on the rate of change of the recorded magnetic signal on the tape, the output doubles for every doubling, or octave change, in frequency. For very high frequencies, the width of the gap is not wide enough, and the output decreases rapidly above a frequency corresponding to half the gap width. The noise level limits the low-frequency response.

PROFESSIONAL VIDEO RECORDING

The use of magnetic tape to record video was innovated by the Ampex Corporation with the introduction of the first commercial video tape recorder in 1956, the Model VR-1000. This machine and subsequent professional machines, such as the Ampex Model VR-2000 which was the industry standard from the mid-1960s to the mid-1970s, use two-inch wide tape moving at a transport speed of 15 in/s.

The record and playback heads are mounted in a two-inch diameter wheel which rotates rapidly at a speed of 240 revolutions per second, thereby creating a writing speed of roughly 1500 in/s. Four heads are mounted on the wheel, thereby leading to the term quadruplex rotary-head assembly to describe the system, or simply "quad" for short. The same heads are used for recording and for playback. The tape cups itself across the wheel as it slowly passes the rapidly rotating heads. The result is a series of tracks that are slightly skewed transversely across the surface of the tape, as shown in Figure 9-6. As one head completes its pass, the video signal is switched to the next head for its pass across the tape. The signal is coupled to the rotating heads either through slip rings or rotating transformers.

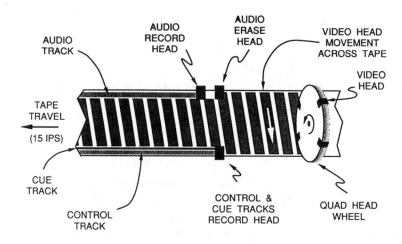

Fig. 9-6 Professional quad-head video recorders use transverse video tracks across the width of the tape. An audio track is recorded at the top of the tape. A control track and a cue track are recorded along the bottom of the tape. Sixteen video tracks contain the information in one video field. A physical space, called a *guard band,* separates each video track to prevent interference. The video tracks are 10 mil wide (10/1000 inch), and the guard bands are 5 mil wide. The transverse video tracks have only a slight tilt. The tape is viewed from the oxide surface. The drawing is not to scale, and the actual video tracks would be thinner and less tilted.

A physical space, called a *guard band,* exists between each track across the tape to prevent interference on playback. The guard bands are 5 mil wide (5/1000 inch), and the video tracks are 10 mil wide. Since the video tracks are recorded across the tape transversely with very little tilt, any longitudinal stretching of the tape will have virtually no time-base error effects on the video signal.

A field consists of 16 tracks, or passes, across the tape. Audio signals are recorded conventionally along the top and bottom edges of the tape. A control track consisting of pulses at the frame rate along with a 240-Hz tone is recorded along the bottom of the tape. The 240-Hz tone is used to control precisely the speed of the rotating heads.

The composite NTSC video signal is used to frequency modulate a high-frequency carrier tone (sine wave). The tip of the sync pulse is placed at 7.06 MHz, and peak white is placed at 10 MHz. The width of the spectrum of the frequency-modulated signal recorded on the tape extends from about 1 to 13.6 MHz for a color television signal. The FM signal is usually recorded at a high enough level to saturate the tape magnetically, since the information is encoded in the frequency variations of the FM signal and not in its amplitude.

In the late 1970s, the quad head was replaced by a series of professional machines that used one-inch wide tape and a helical scanning technique with direct frequency-modulation of the baseband video signal. The tape was wrapped around a drum so that the tape was in the shape of a helix. The record and playback head was mounted in the upper portion of the drum which rotated at a high speed. The result was a series of tracks that slanted across the longitudinal dimension of the tape. The helical scanning technique later was used as the basis for the home video cassette recorder.

HOME VIDEO CASSETTE RECORDING

The video cassette recorder for the home has caused a revolution in the world of video. In addition to its ability to play prerecorded tapes, consumers quickly discovered that programs could be recorded off the air at one time for later viewing at a more convenient time through the use of a timer clock. The time displacement feature of video cassette recorders is an important advantage to many consumers. Another aspect of VCRs was the emergence in every neighborhood of small shops that rent prerecorded tapes on a daily basis for home viewing.

There are two incompatible formats that are used for home video cassette recording. The Betamax (Beta)™ format was developed first and introduced by the Sony Corporation in 1976. The Video Home System (VHS)™ was developed by the Victor Company of Japan and was introduced shortly after the Beta format. The two formats are essentially the same in terms of the basic principles of operation. The differences are in minor details relating to the physical size of the cassettes, the automatic tape loading mechanism, and the precise frequencies used to encode the signals on the tape. The VHS format innovated in offering initially more recording time and also in being less expensive than the Beta format. Hence the VHS format has gained a much larger share of the market than the Beta format.

In video cassette recorders, the signal is recorded along the tape at a slant. This is accomplished by wrapping the tape halfway around a drum at an angle, as shown in Figure 9-7.

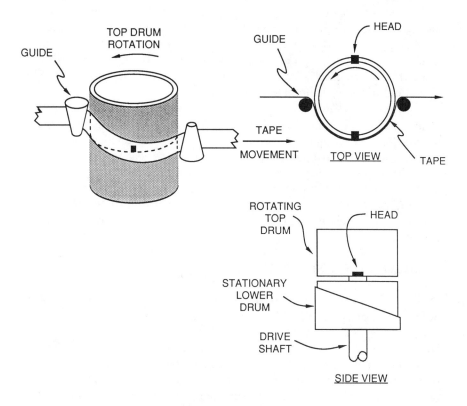

Fig. 9-7 In a home video cassette recorder, the tape is wrapped halfway around a drum. Two rotating heads located 180° apart in the top drum record the video information as a series of slanting tracks across the tape. The tape movement and rotation of the heads are in the same direction. A guide machined into the stationary lower drum assures alignment of the tape around the drum.

The upper portion of the drum rotates at about 30 revolutions per second and contains the record and playback heads. The result is a series of slanted passes along the tape, as depicted in Figure 9-8. The technique is called *helical-scan* or *slant-track* recording. Two heads are mounted

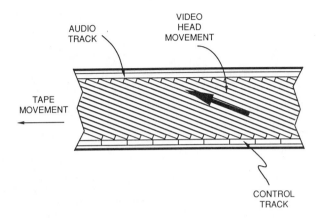

Fig. 9-8 The video information is recorded on gently slanting tracks on the tape. The actual slant of the tracks is much more gentle than drawn here. Pulses at the frame rate are recorded on a control track at the bottom edge of the tape for synchronization purposes and speed control. The audio signal is recorded along the top edge of the tape. The tape is viewed from the oxide surface.

opposite each other on the drum. The heads protrude slightly from the drum to make contact with the surface of the tape. The drum has a diameter of nearly three inches, and the length of each pass along the skewed tape is roughly four and one-half inches. Since the tape has a width of one-half inch, the slant along the tape is indeed quite small.

Sound is recorded conventionally along one or two (for stereo) tracks at the top of the tape. A control track consisting of pulses at the frame rate is recorded conventionally along a track at the bottom of the tape. The tape moves at about 1.5 in/s for normal playing times, and at slower speeds for longer playing times.

The video heads are slightly tilted as they record and play back the signal. This tilt is called *azimuth*. One head is tilted in a direction opposite that of the other head, as shown in Figure 9-9. The result is to minimize interference from adjacent tracks so that the tracks can be recorded immediately adjacent to each other with no physical guard bands. The technique is called a zero guard-band system and greatly increases the length of time that can be recorded on a fixed amount of tape in a cassette.

The video signal is separated into its luminance and chrominance components for recording on home video cassette recorders. Although this results in a compromise in quality, it is necessary for a practical and

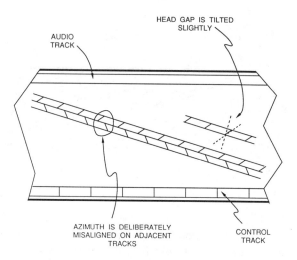

AUDIO
TRACK

HEAD GAP IS TILTED
SLIGHTLY

AZIMUTH IS DELIBERATELY
MISALIGNED ON ADJACENT
TRACKS

CONTROL
TRACK

Fig. 9-9 The heads are tilted slightly in opposite directions with respect to the video tracks. This tilt, called azimuth, decreases the interference between tracks for the luminance signal so that the tracks can be placed immediately adjacent to each other with zero guard band.

affordable home system. The technique is called the *color-under,* or *chroma-under,* system and is shown in Figure 9-10.

The luminance signal has a very wide spectrum ranging from dc, or 0 Hz, to over 4 MHz. This wideband signal could not be recorded directly on a home system. As a first step, the luminance signal is low-pass filtered to about 2.2 MHz. This eliminates some of the higher frequency detail in the picture, and a reduced horizontal resolution of about only 230 lines is obtained.

The instantaneous amplitude variation of the luminance signal is used to modulate the frequency of a pure-tone (sine wave) carrier at about 3.4 MHz. The luminance signal is set so that the tips of the sync pulses cause no modulation of the carrier. The maximum frequency excursion of the carrier occurs at the peak white level of the luminance signal and is fixed at about 1 MHz above the carrier frequency of 3.4 MHz. This type of narrow-band frequency modulation results in a lower sideband extending about 2 MHz below the carrier and an upper sideband extending to about 2 MHz above the carrier.

The chrominance signal is low-pass filtered to 500 kHz. This eliminates some of the spatial detail in the *I* signal. The chrominance signal, also called the chroma signal, is used to modulate the amplitude of a

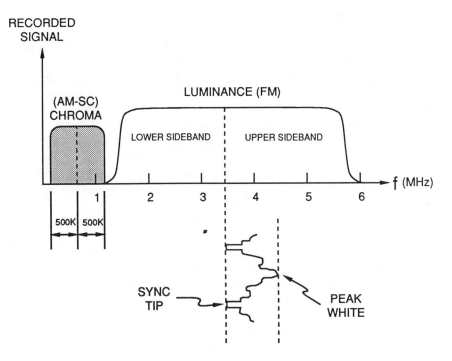

Fig. 9-10 The chrominance, or chroma, signal is separated from the luminance signal and is translated to a lower portion of the spectrum. The luminance signal frequency modulates a sine wave so that the sync tip is at about 3.4 MHz and peak white is at about 4.4 MHz. The chroma signal amplitude modulates (suppressed carrier) a sine wave at about 700 kHz. These two modulated waveforms are added together and are recorded together on the tape. The frequency-modulated luminance signal acts as a high-frequency bias signal for the amplitude-modulated chrominance signal.

carrier at about 690 kHz. The spectrum of the amplitude-modulated carrier extends 500 kHz below and 500 kHz above the carrier frequency. The carrier is suppressed if no chroma signal is present (i.e., suppressed-carrier amplitude modulation is used). The lower sideband of the luminance channel does not extend into the chroma band.

The frequency modulated luminance signal forms the bias signal for the amplitude modulated chrominance signal. This is accomplished by simply adding the two signals.

BETA AND VHS

The Beta format uses a carrier of 688 kHz for the chroma signal, and the VHS format uses 629 kHz. The Beta format uses 3.5 MHz for the sync tips and 4.8 MHz for peak white for the frequency modulated luminance signal. The VHS format uses 3.4 MHz for the sync tips and 4.4 MHz for peak white. The azimuth tilt is $\pm 7°$ for the Beta format and $\pm 6°$ for the VHS format. The drum diameter is 74.5 mm (2.9 inches) for the Beta format and 62.0 mm (2.4 inches) for the VHS format. The Beta format uses heads with a gap of 0.6 μm, and the VHS format uses heads with a gap of 0.3 μm. The Beta format requires the larger diameter drum to achieve higher writing speeds along the tape to compensate for the wider head gap.

The tape in the cassette is automatically threaded around the video-recording drum and other heads in the recorder. The threading mechanism used for the Beta format is a loading ring, while the VHS format uses parallel loading poles in a so-called M threading system. The physical size of the cassettes themselves are also different, although they both have a hinged lid to protect the tape. The Beta cassette is 96 mm (millimeter) wide, 156 mm long, and 25 mm thick. The VHS cassette is slightly larger at 104 mm wide, 188 mm long, and 25 mm thick. Early Beta cassettes contained 150 m of tape for one hour of recording time, and early VHS cassettes contained 247 m of tape for two hours of recording time.

The tape speed in early Beta machines was 40 mm/s for one hour of recording time. The tape speed in the early VHS machines was a slower 33.35 mm/s with more tape in the cassette so that two hours could be recorded. The next generation of Beta machines operated at 20 mm/s, thereby doubling the recording time to two hours. The VHS machines likewise halved their time to 16.67 mm/s, thereby responding with four hours of recording time. The slower tape speeds require thinner video tracks which require thinner record heads. The pair of thinner heads are mounted on the drum opposite the other pair of heads so that the machine has four heads. The signal-to-noise ratio is worse with thinner heads, so some loss in quality is necessary to obtain the longer recording times.

The horizontal resolution possible with VHS can be estimated as follows. The chrominance signal occupies a spectrum space centered at 629 kHz with a width 500 kHz above and below the center frequency. Thus, the upper frequency limit on the chrominance band is $629 + 500 = 1,129$ kHz, or about 1.2 MHz. The luminance signal frequency modulates a carrier of 3.4 MHz. There thus is a space of

3.4 − 1.2 = 2.2 MHz available for the lower sideband of the frequency-modulated luminance carrier. Since the spectrum of a modulated carrier is symmetric, the total bandwidth of the modulated carrier is 4.4 MHz.

The ratio of the peak deviation to the maximum frequency in the signal is called the index of modulation. For the small indices of modulation used in VHS, the bandwidth of the frequency-modulated signal is the same as that of amplitude modulation, or twice the maximum frequency in the modulating signal. Thus, only about 2.2 MHz is available for the actual luminance signal.

The video portion of a horizontal scan line requires about 53 μs. Multiplying 2.2 MHz by 53 μs gives 116 as the maximum number of cycles that are possible in a scan line. Thus, twice this number, or about 230 vertical lines, could be resolved in a scan line. The horizontal resolution, therefore, is about 230 lines.

One feature offered in many VCRs is stop motion, or still frame. For stop motion, the tape is stationary, and the heads in the drum continuously scan the same track. The problem is that when the track was recorded the tape was moving and this forward motion of the tape contributed to the slant of the track. For stop motion, the tape is no longer moving, and hence the heads do not follow the same slant as when the moving tape was recorded. The result is mistracking, the heads spill over into an adjacent track, sync is momentarily lost, and the picture tears a little. Newer VCRs have an additional head that is positioned to minimize this problem.

The home video cassette recorder has its own front-end tuner to receive particular television RF channels. The audio, luminance, and chrominance signals are separated and RF demodulated in preparation for recording. The chrominance signal is not separated into its I and Q components, but is simply translated down in frequency and recorded as a single signal. Upon playback, the luminance and chrominance signals are combined to create a single NTSC video signal. The video and audio signals are then used to modulate RF carriers for either channel 3 or 4, depending which channel is vacant in the area. The RF output of the VCR is connected to the antenna input of the television receiver for viewing on either channel 3 or 4.

A new video cassette about the size of an audio cassette using a tape with a width of 8 mm (about one-third of an inch) has been introduced. The cassette is used in a single portable unit that combines the camera and the video recorder and weighs only five pounds. The new unit is called a "camcorder." A new tape with superfine particles was developed for this application. The recorder has a smaller diameter drum, and the tape is wound nearly all the way around the drum.

SUPER VHS

The Victor Company of Japan (JVC), an independent subsidiary of Matsushita Electric, has developed an improved form of its VHS format called Super VHS™, or S-VHS for short. An improved tape is used so that a broader bandwidth can be recorded on the tape. The luminance signal is recorded using frequency modulation in fashion identical to the VHS format, except that the S-VHS format uses 5.4 MHz for the sync tips and 7.0 MHz for peak white. Thus the frequency deviation has been increased from the 1.0 MHz used for VHS to 1.6 MHz for S-VHS.

A carrier of 629 kHz is used for the chrominance signal, and the upper frequency limit on the chrominance band is about 1.2 MHz. The increase in the frequency of the FM carrier from the 3.4 MHz used for VHS to the 5.4 MHz used for S-VHS thus results in an available bandwidth of 4.2 MHz for the luminance signal. This means that a horizontal resolution of 430 lines is possible with S-VHS.

The S-VHS machines offer separate video outputs for the luminance *(Y)* signal and the chrominance *(C)* signal. If an appropriate television monitor is used, the result is a considerable improvement in the quality of the image, since the color distortions that would result from a combination of the *Y* and *C* signals to create a standard composite NTSC video signal have been avoided.

TIME-BASE ERROR AND OTHER PROBLEMS

The slant recording technique allows the very high writing speeds required to record high frequencies, while at the same time enabling a slow tape speed. The video track, however, is recorded along the long length of the tape which is subject to varying tensions, stretching, and vibration while being recorded or played back. Also, the heads are never precisely 180° apart on the drum. The result is a frequency instability, or jitter, that is technically called *time-base error*. If not corrected, the final result would be constant changes and errors in the hue of the color in the picture.

Time-base error is corrected through the use of a signal locally generated in the VCR that is varying in exact synchrony with the time-base errors in the luminance signal. This locally generated signal is mixed with the luminance signal, and the jitter cancels in the frequency-difference signal. Perhaps this can be made more understandable by an example using the Beta format. The chrominance signal is at 0.688 MHz on the tape. On playback, the result is a jittered chrominance signal, say now at 0.688 + β MHz, where β represents the jitter. The jittered chrominance signal is mixed with a local source of 3.58 MHz, and the sum-frequency signal resulting from the mixing process has a frequency of 4.27 + β MHz. This signal is then mixed with the jittered chrominance signal. The jitter β cancels in the difference-frequency signal, thereby leaving an unjittered, or time-base error corrected, luminance signal at 3.58 MHz.

One advantage of the color-under system used in VCRs is that the chrominance signal does not need to be demodulated for recording. The chrominance signal simply is moved from 3.58 MHz down to 0.688 MHz or 0.629 MHz, depending upon whether the Beta or VHS format is used. A problem, however, is that at this low frequency the tilted-azimuth technique is not as effective at preventing interference, or crosstalk, from adjacent tracks as it is for the higher frequency luminance signal. This problem, called color crosstalk, is solved by the use of a special filter, called a comb filter, in combination with phase reversals of the signals recorded by each head. The particulars of the technique are somewhat too involved to be explained here.

References

Athey, Skipwith W., *Magnetic Tape Recording,* National Aeronautics and Space Administration (Washington, DC), 1966.

Ginsburg, Charles P., "Comprehensive Description of the Ampex Video Tape Recorder," *J. SMPTE,* Vol. 66, April 1957, pp. 177–184.

Jones, E. Hayden, *Audio Frequency Engineering,* Chatto and Windus (London), 1961, pp. 174–190.

Kihara, Nobutoshi, and Fumio Kohno, "Development of a New System of Cassette Type Consumer VTR," IEEE Trans. Consum. Electron., Vol. CE-22, February 1976, pp. 26–35.

Kybett, Harry, *Video Tape Recorders,* Howard W. Sams (Indianapolis), 1978.

McComb, Gordon, "Super Pictures From Super-VHS," *Popular Science,* January 1988, pp. 68–70 and 114.

McGinty, Gerald P., *Videocassette Recorders: Theory and Servicing,* McGraw-Hill (New York), 1979.

Olson, Harry F., *Acoustical Engineering,* Van Nostrand (Princeton, NJ), 1957, pp. 384–390.

Shiraishi, Yuma, and Akira Hirota, "Video Cassette Recorder Development for Consumers," *IEEE Trans. Consum. Electron.,* Vol. CE-24, August 1978, pp. 468–472.

10. Video Discs

INTRODUCTION

The video disc was certainly heralded with great expectations. In 1979, the President of RCA predicted a "30 to 50 percent penetration of all U.S. color television homes in ten years" (RCA Quarterly Review for Shareholders, Fourth Quarter 1979, p. 2). Well, things sometimes do not turn out as expected! RCA's video disc for the home was withdrawn five years later because of an overall lack of acceptance by consumers, and it now is commercially dead.

The RCA video disc was under development since 1964, was introduced into the marketplace in 1981, and was withdrawn three years later after sustaining losses well over one-half billion dollars (*Information Display*, February 1982, p. 6; *Time*, April 16, 1982, p. 47). An optical video disc developed by Phillips did not fare much better. RCA was able to sell over 100,000 video disc players in the first nine months, and Phillips sold about 75,000 of its players (*The Economist*, March 20, 1982, p. 93). These sales were far below expectation and led to the market failures of both systems.

The basic idea of the video disc was that it would be a video "phonograph" disc. All the convenience and familiarity of the black-vinyl phonograph disc would supposedly transfer to the video disc, resulting in a large market for movies and other prerecorded video programming in the home. One problem was, however, that two noncompatible technologies were introduced for video discs; one used a laser and the other used a mechanical stylus to read the video information recorded on the disc. These two different approaches created confusion for many consumers.

Another problem was that the disc was a play-only medium and could not be used for recording. The competing technology was the video tape recorder using fairly convenient cassettes. The video disc took long to bring to market, and, when it finally arrived, it was far too late, since the video cassette recorder had firmly entrenched itself. Although the video disc produced a superior quality television picture, most television sets did not have the inherent quality to show the difference. Furthermore, many consumers had discovered the advantage of the VCR for unattended recording of a television program for more convenient viewing at a later time. This, coupled with the ability to rent VCR movies at neighborhood stores, was just too much for the video disc to overcome and led to its death in the marketplace.

The death of a product can be a sad occasion, particularly for its inventors and developers. In the case of the video disc, the laser approach was able to resurrect itself in the form of a much smaller disc on which the large bandwidth was used to record digital audio. Thus, from the ashes of the laser video disc arose the digital audio compact disc — a phenomenal success in the marketplace.

CAPACITANCE ELECTRONIC DISC (CED)

Two different technologies were invented for video discs. One technology used a laser beam to read the video information stored on the disc, and the other technology used a mechanical stylus to play the video information. With both approaches, the diameter of the disc was twelve inches and both sides of the disc contained video information.

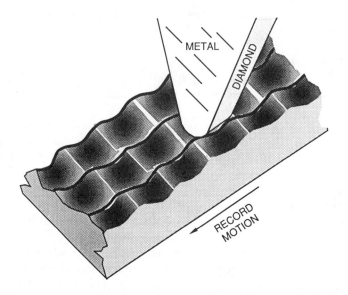

Fig. 10-1 The spiral grooves in the RCA SelectaVision video disc are read by a diamond stylus with a metal coating on its front surface. The stylus is guided along the grooves by their shallow-V cross-sectional shape. The grooves contain varying up-and-down, or hill-and-dale, undulations, and the video signal is encoded by the variations in these undulations.

The mechanical stylus approach invented by RCA and marketed under the name SelectaVision™ is based on a *capacitance electronic disc* (CED). The video information is read from the disc by a diamond stylus in a fashion vaguely similar to the way a phonograph record is played.

The video information is encoded on the surface of the disc as the up-and-down, or hill-and-dale, undulation of a continuous spiral groove with a shallow V-shaped cross section, as shown in Figure 10-1. The stylus is mechanically forced to track the spiral because of its physical grooved structure. The disc turns in a clockwise direction and is read from the outside to the inside.

The stylus rides across the hill-and-dale undulations in the groove. The video information is read from the undulations in terms of electrical effects caused by the varying distance between the stylus and the bottom of the groove, as is explained in the next paragraph.

The surface of the disc is a sandwich consisting of a bottom metal layer (usually aluminum), a layer of plastic dielectric, and a top layer of lubricant, as shown in Figure 10-2. The front surface of the diamond stylus is coated with metal. As the groove moves beneath the stylus, a varying electrical capacitance is created between the metallic coating on the front of the stylus and the bottom metal layer on the disc. The frequency of a tuned resonant circuit is caused to change by this varying capacitance, and this results in a frequency-modulated signal which can be demodulated to obtain the composite video signal.

The video and audio television signals are recorded on the disc in the following fashion. The video signal frequency modulates a high-frequency subcarrier with the sync tips at 4.3 MHz and peak white at 6.3 MHz. The audio signal frequency modulates a subcarrier at 716 kHz. These two frequency modulated subcarriers are added together to create the signal that is recorded onto the disc. Since the bandwidth available for the video signal is only 3 MHz, the color subcarrier used for the chrominance signal recorded on the disc is 1.53 MHz, rather than the 3.57 MHz color subcarrier used for conventional NTSC color television.

The CED 12-inch diameter disc turns at 450 revolutions per minute (rpm), and one side contains 60 minutes of video. At this rate of revolution, each circular groove contains four one-thirtieth of a second television frames. There are over 5,000 grooves per inch on a CED disc, as compared to the nearly 300 grooves per inch of a phonograph record.

The surface of the disc needs to be protected, and hence the disc is stored in a protective plastic sleeve or caddy from which it is automatically extracted for play by a mechanism in the player. Since there is mechanical contact of the stylus with the surface of the disc, both the stylus and the disc are subject to wear with a resulting deterioration in picture quality.

The capacitance disc (CED) was introduced into the marketplace by RCA in 1981. The player retailed for $500, and the discs sold for $20. The product was withdrawn in 1984.

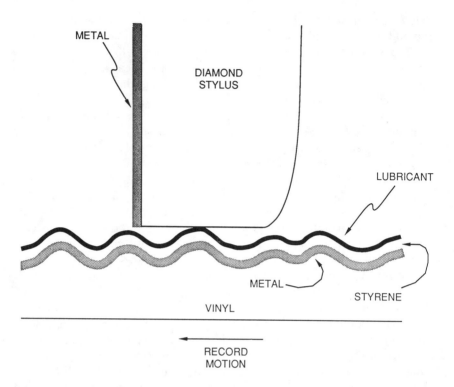

Fig. 10-2 Cross-sectional view of the sandwich-like composition of the RCA video disc. As the record surface moves under the stylus, a varying capacitance is created between the metal on the front surface of the stylus and the metal coating slightly below the surface of the disc. The system is called a capacitance electronic disc (CED).

A CED, called a *video high-density disc* (VHD), was developed by the Victor Company of Japan (JVC), an independent subsidiary of Matsushita Electric. The VHD is essentially similar to the RCA product in terms of its use of the basic principle of variable-capacitance operation. The VHD rotates at a higher rate of 900 revolutions per minute so that two frames are contained in one revolution. The VHD does not have grooves to guide the stylus, but, instead, a servomechanism is used to guide the stylus as it tracks the spiral of video information stored on the disc.

OPTICAL VIDEO DISC

The wear problems of the CED are avoided with the *optical video disc* in which a laser beam of light reads the video information stored on the disc with no mechanical contact with its surface. The optical disc, invented by the firm of N. V. Phillips in The Netherlands, was marketed in the United States jointly with the Music Corporation of America (MCA) under the name LaserVision™.

The video signal on a video optical disc is encoded as a spiral of reflective microscopic pits, as shown in Figures 10-3 and 10-4. The pits are protected by a layer of hard transparent plastic. A *laser* beam is focused on the pits and the space between them. The beam is reflected back from the space between the pits but is not reflected back from the pits themselves. The length of the pits is varied with the video signal, and this information is read from the disc as the length of time that the laser beam is not reflected back from the disc. The pits and the space between them are coated with an aluminum reflective coating. The pits are 0.4 μm wide, and the spirals are spaced 1.6 μm apart.

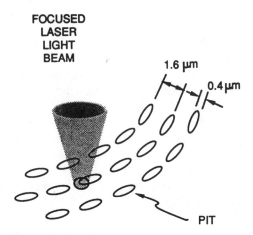

Fig. 10-3 The surface of a laser video disc consists of a spiral composed of pits that vary in length. The varying length of the pits encodes the video signal, and the information is read from the surface of the disc by a focused beam of laser light. The pits are 0.4 μm wide, and the spirals are spaced 1.6 μm apart.

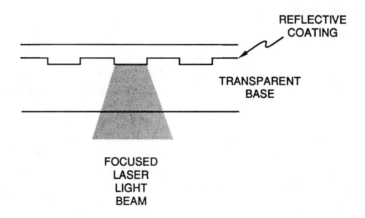

FOCUSED
LASER
LIGHT
BEAM

Fig. 10-4 Cross-sectional view of a laser video disc. The pits and space between them are covered with a reflective coating. The laser beam cancels itself when reading a pit but is reflected back from the space between the pits. The laser beam is highly focused so that dirt and imperfections on the surface of the disc are ignored.

The technique used to encode the video signal is somewhat similar to the FM used for video cassette recorders. The audio signal frequency modulates a carrier at 2.3 MHz. The composite color video signal frequency modulates a carrier at 7.6 MHz. The sync tips are at 7.6 MHz, and maximum white is at 9.3 MHz. The maximum frequency deviation thus is 1.7 MHz. The audio FM signal and the video FM signal are added together, as depicted in Figure 10-5. The zero-axis crossings of this sum signal are used to determine the length of the pits recorded in the spiral on the optical disc. With this form of pulsewidth modulation, the length of the pits gives information about the audio signal and the rate at which the pits occur gives information about the video signal. A second audio channel frequency modulates a carrier at 2.8 MHz and is added to the first channel at 2.3 MHz.

The laser beam is highly focused on the pits and the space between them. Any dirt, scratches, or other blemishes on the outer surface of the disc is out of focus and hence has little effect on the laser signal reflected from the disc. In effect, the laser beam "sees" through the blemishes.

The laser itself, along with the sensor for receiving the reflected light signal, is mounted on an assembly that can slide across the disc to track the spiral. Minute tracking errors are corrected by mirrors.

The laser beam is optically separated into three beams that scan the disc. The central beam reads the encoded signal. The two outer beams

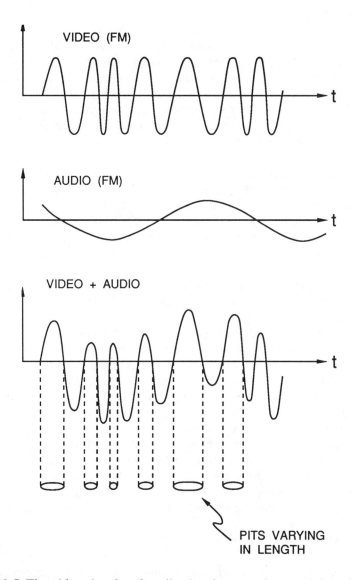

Fig. 10-5 The video signal and audio signal are used to modulate the frequencies of two separate carriers. The two frequency-modulated carriers are then added together. The zero-axis crossings of this signal determine the lengths of the pits recorded on the laser video disc.

are used to track the spiral of information as the disc rotates. The disc rotates in a clockwise direction and is read from the inside to the outside.

A *servomechanism* compares a measured signal with some reference and then creates an adjustment signal to adjust appropriately the measured signal to reduce the error. A number of servomechanisms are used in an optical disc system. One servomechanism tracks the spiral and moves the laser beam across the disc. Another servomechanism keeps the beam in focus on the surface of the disc by continuously adjusting an objective lens, and a third mechanism adjusts the speed of the disc to synchronize it with a master clock in the player.

Laser light has a single, precise frequency; it is *monochromatic light*. Laser light has an intense, highly focused beam in which all of the light rays are parallel to each other; it is a *collimated beam*. All of the waves of laser light are in phase with each other; it is *coherent light*. These three properties of laser light are essential for the optical disc. The depth of the pits is precisely equal to one-quarter of the wavelength ($\lambda/4$) of the light. Thus, the laser light in traveling down the pit and back is exactly $180°$ out of phase with the light at the surface or top of the pits. The pit-reflected beam thus cancels the light reflected from the space between the pits.

There are two approaches to optical discs. In the first approach, the disc rotates at a *constant angular velocity* (CAV) of 1,800 rpm. With the CAV approach, one revolution of the disc contains exactly one frame of video. Thus, simply halting the radial movement of the laser beam gives a freeze-frame feature. It is also very simple to access any desired frame. The total time per side, however, is limited to only 30 minutes.

In the second approach, the disc rotates at a *constant linear velocity* (CLV) varying from 1,800 rpm at the inside to 600 rpm at the outside. This approach allows more information to be stored on the disc, and the playing time is 60 minutes per side. Since each revolution no longer contains exactly one frame, however, the freeze frame feature is not so easily available, and it is much more difficult to access a single specified frame.

INTERACTIVE VIDEO DISC

The CAV video disc has been coupled with computer technology to create an interactive storage medium for either video information or digital data. In the video interactive system, information to identify single video frames is placed in the *vertical blanking interval* (VBI). A digital computer causes the disc player to search for a specified frame. Individual frames can be accessed and displayed. Also, a specific frame can be accessed and motion video continued from there to a specified frame at which motion video should stop.

The interactive video disc is particularly useful where individual frames and short stretches of motion video need to be accessed and displayed. The applications are mostly in education and training. Questions might be displayed on the screen and then, depending upon the answer, different segments of video might be shown.

Although the interactive video disc appears quite exciting for education and training, its use is not extensive, and the technology is still costly. One is left wondering whether, despite all its promise, this is not just another product to be added to the closet of unused educational technology.

The optical disc has a very large bandwidth and thus is quite appropriate for storing large quantities of digital data. The on-off patterns of a digital signal are particularly appropriate for the reflected or not reflected signal produced by the laser beam as it reads the information stored on the disc. One obvious application is to use the optical disc to store digital audio, and the $4\frac{3}{4}$-inch diameter compact disc that has conquered the audio world does exactly that.

Another application is to store digital data that represents eight-bit ASCII coded text. Used this way, a 12-inch optical disc could store all the information in the *Encyclopaedia Britannica*. Other possible applications would be to store all the information contained in large catalogs, parts lists, and directories.

References

Brandinger, Jay J., "Six Million Video Discs in 1982," *Information Display,* February 1982, pp. 3–6.

Clemens, Jon K., "Video Discs: Three Choices," *IEEE Spectrum,* March 1982, pp. 38–42.

Fox, Barry, "Video Discs — Too Late for the Gravy Train?," *New Scientist,* July 30, 1981, pp. 277–280.

Free, John R., "Video Discs for Your Color TV," *Popular Science,* November 1974, pp. 92–95 and 144–145.

Free, John R., "Video-Disc Players," *Popular Science,* February 1977, pp. 85–87, 140.

Free, John R., "Optical Disc Can Store an Encyclopedia," *Popular Science,* August 1982, pp. 47–50.

Koepp, Stephen, "Slipped Disc," *Time,* April 16, 1984, p. 47.

Mennie, Don, "Television on a Silver Platter," *IEEE Spectrum,* August 1975, pp. 34–39.

11. CATV and Alternative Broadcast Technology

INTRODUCTION

Cable television (CATV) uses a coaxial cable to carry a variety of television channels. The television signals are obtained from a variety of sources. The most popular source is over-the-air broadcast VHF and UHF television. Another source is satellite stations that exist solely to provide television programming to CATV operators. These satellite stations broadcast a variety of programming including an all-news network operated by Ted Turner, movies offered by Home Box Office, religious services and sermons, special sports events, and rock videos. A CATV operator might also use a video tape recorder to provide further program material to send over the CATV system. Some CATV systems include their own television studio from which local community shows originate.

All the programming materials from all these sources are combined to cover a fairly large frequency spectrum and are then transmitted over the coaxial cable network. Amplifiers are used in the network to amplify the signal along the cable. A decoder is used in the home to choose the desired cable channel and then to display it on the home television receiver. A CATV system is, indeed, very simple, conceptually and technically.

The creativity in CATV involves the marketing of the service to the public. A selection of programs needs to be packaged at a price the public is willing to pay on a monthly basis. Different packages are frequently offered at different prices. The most basic package usually consists of the conventional broadcast UHF and VHF stations.

CATV had its beginnings in 1958. At that time, television was in its infancy and many areas of the country were not served by television. A solution was to erect a large television antenna on a hilltop to receive the distant television signals. These signals were then amplified and distributed to homes in the community over coaxial cables. The system was called *community antenna television,* or CATV for short. CATV was the only way many communities could have television, because they were far from large cities, or because the intervening terrain was hilly, thereby causing poor over-the-air reception. These community antenna systems usually offered twelve or fewer channels.

These early CATV systems were "mom-and-pop" businesses, frequently operated by the local TV store in town. CATV today is a big business in which many local CATV systems are owned by such large *multiple-system operators* (MSOs) as Warner Amex Cable Communications, Group W Cable, Cox Communications, and Tele-Communications (TCI). Many of these MSOs are owned by such multimedia firms as Time, Inc., Newhouse Communications, and the Times Mirror Company.

During these early days of CATV, the broadcasters viewed CATV in a positive light, since the audience receiving their programs — and commercial advertising — was increased. Some CATV operators, however, started to use microwave technology to import television signals from distant areas, thereby creating viewing competition with the local stations. In this way, CATV was being used in areas with good over-the-air reception as a means to increase program variety. CATV thus came to mean cable television in a much broader sense than its community-antenna beginnings. In large cities, CATV was essential to avoid the ghosts caused by multiple reflections of the over-the-air signals from many large buildings. Later, cable systems were developed that offered many more channels for viewing, thereby increasing the variety of programming for the subscriber. Some newer cable systems also offer special events for which the subscriber must pay on a per-event basis.

The importing of distant signals from one television market into another attracted the attention of the television networks and their affiliated stations. Broadcasters thus became concerned about CATV, and the FCC was encouraged to exert its regulatory influence over CATV.

In 1966, the FCC issued rules that forbade CATV operators in the one hundred largest television markets from importing distant television signals. The authority of the FCC to regulate CATV was challenged by the CATV industry, but was upheld by the United States Supreme Court in 1968. The FCC then issued rules that actually halted further expansion of CATV in the one hundred largest markets. In 1972, this ban was lifted, but the FCC then issued new rules that controlled the type and content of programs that could be offered on CATV. In 1977, a federal appeals court ordered the FCC to repeal these restrictions.

In some respects, the public considers broadcast television to be a free good. If one is located far from a large city, then CATV is clearly viewed positively, since it is the only way to obtain television signals. If one is near a large television market, however, then CATV is viewed as an attempt to charge for television signals that can be obtained for free from over-the-air broadcast stations. I can remember a time when "pay TV" was actually thought by some folks to be a communist conspiracy!

At present, the public has a tremendous variety of sources of television programs, and CATV is only one of these many sources. Video cassette rentals have quickly become a source of movie viewing in the home. Over-the-air television broadcasting by the major networks and their affiliates is still alive, but is facing ever increasing competition and declining market share. The over-the-air VHF spectrum is reasonably full in most television markets, but plenty of UHF space is still available. The technology is available to allow low-power VHF stations in some markets to serve specific community needs for specialized programs. The video disc failed in the marketplace, but the video cassette has become a great success.

In the early 1970s, the CATV operators dreamed of the wiring of the nation with coaxial cable for television signals and a variety of other poorly defined futuristic two-way communication services. The vision of great profit led to battle among the corporate giants of the CATV industry to obtain local franchises for CATV systems in most major markets. These franchises were obtained at considerable cost, but the market stabilized soon afterward. The new two-way services failed to materialize. The wide variety of alternative sources of television programming led to a public apathy toward CATV. The rise of satellite networks to provide the CATV operators with program material has been exciting, but even that segment of the industry has now stabilized. Cable has become just another source of television.

CATV SYSTEM

A modern CATV system begins at the *headend,* where the television signals originate for transmission *via* a *cable distribution network.* The headend receives over-the-air VHF and UHF television signals for re-transmission to its subscribers. The headend also receives satellite signals, which originate from studios that supply movies, news, weather, and other broadcast services to the cable companies for a fee. Television program material is also obtained *via* private microwave links from distant markets. The headend might also obtain video material for its own tape playback or from its own studio. The latter are called *local originations.* Figure 11-1 shows a schematic of a typical CATV system.

In addition to the headend, a CATV system also consists of a coaxial-cable network to distribute the signals within a community. The television programs received and originated at the headend are offset in frequency and are combined for retransmission over the cable network.

The cable network looks like a tree with many branches extending from a "trunk." The signal originates at the base of the tree, the headend, and is carried along the trunk to the "branches" and finally to the "leaves." A cable network consists of a *trunk system, distribution lines* (the branches), and *subscriber drops* (the leaves). The coaxial cable used in a CATV system has an inner conductor surrounded by an outer conductor. The

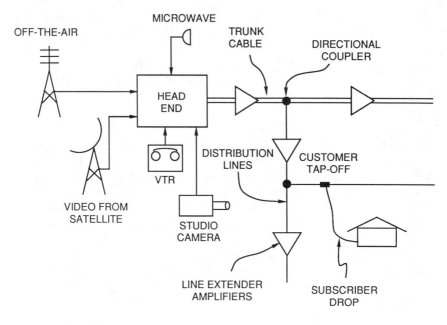

Fig. 11-1 Schematic representation of a typical CATV system. The programs are transmitted from the headend to the homes of the subscribers *via* a coaxial cable network.

two conductors are electrically insulated from each other either by a plastic foam, or sometimes the coaxial cable used in the trunk system consists of a hollow cable in which plastic spacers separate the inner conductor from the outer conductor. Coaxial cable has a very large bandwidth extending up to 400 MHz and, thus, can carry a large number of television channels. The coaxial cable used in CATV has a characteristic impedance of 75 Ω.

The trunk system forms the backbone distribution for the cable network. From three to four trunks will branch out from the headend in a typical cable system. The trunk system consists of coaxial cable with an outside diameter of one inch to three-quarters of an inch.

The signal along the cable continuously loses its strength because of electrical resistive losses in the cable and thus needs to be amplified. These losses are overcome through the use of amplifiers placed every 1,400 to 2,000 feet along the trunk cable. The amplifiers increase the input signal by about 20 dB, which is equivalent to an increase by a factor of ten in voltage. The cable does not pass all frequencies equally well, and different losses are encountered for different channels on the cable. This effect is neutralized through the use of filters, called *equalizers,* that are used along with the amplifiers to ensure that the electrical level is the same for all channels.

The distribution lines branch off from the trunk. A small amount of the signal on the trunk is siphoned from the trunk through the use of a *directional coupler.* The directional coupler uses magnetic and capacitive induction to tap off the signal from the trunk. Any signal that is reflected back along the trunk cable is ignored by the directional coupler, hence its name *directional.* The output of the directional coupler is about 13 dB less than the signal level on the trunk cable. This loss is overcome through the use of a *bridging amplifier* at the directional coupler. The bridging amplifier electrically isolates the trunk from the distribution line and also generates a strong electrical signal to be fed down the line. Typical bridging amplifiers have from 20 to 40 dB gain.

The distribution lines carry the television signals within a subscriber neighborhood. The distribution lines use coaxial cable with an outside diameter of from 0.4 to 0.5 inch. If the distribution line is particularly long, amplifiers called *line-extender amplifiers* are used to increase the signal level about 20 to 30 dB.

The CATV connection to a subscriber home is made through a 0.25-inch diameter coaxial cable called a subscriber drop. The subscriber drop is connected to the distribution line by a *customer tap off.* As many as four subscriber drops will be served by a single customer tap off. Directional couplers are usually used as subscriber tap offs. The electrical isolation given by these couplers prevents any harm from occurring to the distribution signal that might be caused by problems within the home of a subscriber.

The television channels on the cable are transmitted at different frequencies than those used for over-the-air broadcast transmission. Thus, the cable channels need to be converted in frequency to frequencies that can be received on the home television receiver. This conversion is accomplished by a converter supplied by the CATV operator. The converter usually has a tuning mechanism to select a particular cable channel for viewing. The frequency band of the cable channel is then converted to either channel 3 or 4 for viewing on the home television receiver. Some newer television receivers are "cable ready" in the sense that the converter electronics are built into the set. Of course, if the CATV operator scrambles some channels, then a special box will still be required.

The actual signal level in a CATV system is about 100 millivolts maximum along the trunk. The final signal level at the home is about 10 millivolts.

The cable itself is usually installed on local telephone poles, in which case a fee is usually paid to the local telephone company or power utility. The cable can also be placed underground, either in conduit or buried directly. Aerial installation is the least expensive, but local ordinances might require underground installation of the cable.

The various amplifiers in the cable distribution system are powered by direct current at 24 volts. The coaxial cable itself sometimes carries the dc power required by the amplifiers. Alternatively, the amplifiers might be powered by connection to the nearby ac power line.

The coaxial cable used in early CATV systems had enough bandwidth for only a dozen television channels. Newer cable has a bandwidth of 400 MHz or more. The newer cable makes possible CATV service that can offer 100 television channels. The issue then becomes what programming to offer over such a large-capacity medium. Such a large capacity would make it difficult even to prepare a paper-based listing of the programs available each week. One type of novel service to be offered over such large-capacity CATV systems would be continuously to repeat network programs each evening. In that way, the subscribers could catch their favorite programs at virtually any time. Such time-displaced viewing is already possible, however, with the home video cassette recorder.

TWO-WAY CATV

Coaxial cable is a broadband medium capable of carrying not only television signals but also a wide variety of such other signals as voice telephone speech and digital data. Although coaxial cable as used in CATV systems is one-way, it can be made two-way through the addition of appropriate electronic equipment. At one time, some futurists foresaw a nation wired with two-way coaxial cable to each and every home. The concept of the "wired nation" was thus born. Such two-way CATV has been proposed as a means of wiring the nation, although some people have correctly pointed out that the nation is already wired by the two-way telephone network.

The futurists saw the bundling together of a wide variety of services all delivered over the cable. Some of the "new" services proposed for two-way CATV include fire and burglar alarm service, viewdata or videotex information services, and pay-per-event television. Most of the so-called "new" services are already possible using the existing telephone network, and, hence, not surprisingly, two-way CATV has remained a fad.

The transmission of television signals over the cable is a one-way affair. The television signals are transmitted in one direction down the cable from the headend to the homes of the subscribers. Such services as voice telephone service and data communication, however, are two-way and, therefore, require an electrical communication path back from the homes of the subscribers to the headend. This two-way capability is not easily obtained with CATV technology.

The coaxial cable itself is nothing more than a transmission medium. The medium can be divided into frequency bands, for example, in which some bands are used for communication in one direction and other bands are used for communication in the other direction. Thus, in theory, the use of cable as a two-way communication medium is conceptually simple. Unfortunately, the situation is not actually that simple. The CATV distribution system uses amplifiers to amplify the signal every 2,000 feet or thereabouts. Amplifiers are inherently one-way, and thus the CATV system is also inherently one-way. A CATV system can be made two-way, but only through the addition of costly new equipment at each amplifier to create a return bypass path back to the headend.

In response to the euphoria over the prospects for a two-way capability on CATV, in 1972 the FCC issued a rule requiring a two-way capability on all new CATV systems larger than some minimum size. The cable companies quickly promised all sorts of new two-way capabilities and services to win their franchises. Frequently the two-way capability was nothing more than additional space in a rack of electronic equipment so that the two-way electronics could be added at some future date. The FCC's two-way requirement was disbanded in 1979 by the United States Supreme Court.

The earliest CATV system actually to provide a two-way capability was introduced in 1977 by the Warner Amex Cable Communications Company in Columbus, Ohio. This two-way CATV service was called Qube. It offered an interactive capability using a low-speed data path back to the headend. The return path was used for such services as polling and remote ordering of goods. The subscribers were apathetic toward the services offered by Qube, and, after sustaining large losses, Warner greatly curtailed its use of the interactive service.

PAY PER VIEW

There is one interactive service that would appear to make considerable sense for a CATV system. That interactive service is the ability to authorize and pay for a specific cable program as it is being viewed. The subscriber would send a data signal to the headend authorizing payment to view some specific program. The headend would then send an encoded data signal back to the equipment in the home, which would descramble the program to be viewed.

It would seem to make sense to use the cable itself for the return data signal for pay-per-view service. The addition of the capability for a return data path back up the cable, however, would be costly. One alternative, suggested by Gary Schober of New York University's Interactive Telecommunications Program, would be to transmit all of the data signals down the cable to its termination. A special receiver would collect the data signals and then transmit them all back to the headend over a telephone line. Another alternative would be to use the telephone line in the home to place a data call to the headend, where a special data signal would be sent back over the cable to enable descrambling of the television signal.

The technology for pay-per-view service would be a costly addition to the CATV system. The real issue, however, is whether subscribers want such a service. Also, the competition from video cassette rentals would be very keen.

BROADBAND CABLE

The use of broadband amplifiers makes it possible to offer CATV with 100 or more channels. What all these channels will broadcast is another question. Perhaps such 100-channel CATV systems will become like radio in which one scans the dial until something interesting is found, or perhaps the same material will be broadcast over and over so that the viewer can catch a favorite show at any time.

OPTICAL FIBER

Some newer CATV systems are using optical fiber for the trunk system. Optical fiber is an optically transparent thin fiber of glass that guides a beam of light from one of its ends to the other. A solid-state laser is used as the light source at the transmitting end, and a photodetector is used to detect the light signal at the receiving end. Optical fiber has a very low loss, thereby eliminating the need for amplification every couple of thousand feet, as is required with coaxial cable. Optical fiber is not affected by eletromagnetic energy and, thus, is interference free.

Optical fiber is usually used as a digital medium in which pulses of light are turned on and off to encode the amplitude fluctuations of the analog signals carried on the fiber. This means that analog-to-digital and digital-to-analog converters are needed for an optical fiber system. Optical fiber has a very large bandwidth, and hundreds of television signals can be transmitted easily over a single fiber.

The telephone industry is contemplating the use of optical fiber as a replacement for the twisted pairs of copper wire that currently constitute the local loop between the home and the central office. This very same fiber, if used solely for telephone conversations, would be a waste of the tremendous bandwidth of the fiber. One other obvious use would be the transmission of television signals to the home. Although technically feasible, the use of fiber owned by the telephone company to carry television signals would raise many policy and regulatory issues. One possible solution would be to restrict the telephone company to providing only a transmission capability, and the packaging and provision of the television programming would involve the CATV company. It will be interesting to watch how these issues evolve.

DIRECT BROADCAST SATELLITE (DBS)

A communication satellite is a microwave repeater located 22,300 miles above the surface of the earth. At that distance, the satellite makes one complete revolution around the earth in exactly 24 hours. If the orbit of the satellite is directly above the equator of the earth, the satellite will revolve about the earth at exactly the same speed as the earth turns beneath it. The satellite will thus appear stationary with respect to the surface of the earth, or simply suspended in the sky. Such an orbit is called a *geostationary orbit*.

A communication satellite receives a signal transmitted to it from the earth and then simply retransmits the signal back to the earth. The signal is amplified and translated in frequency at the satellite. Communication satellites are used by television networks to distribute television signals around the world and to distribute signals to their local affiliates. Private networks distribute signals to CATV operators.

The signals retransmitted from the satellite can be received by anyone with an appropriate antenna and decoder. In fact, many people, particularly in rural areas, have large antennas, called dishes, that receive the signals sent by the networks to CATV operators and local stations.

It has been proposed to use a high-power transmitter on a communication satellite to cover a whole country. In this way, one master broadcast antenna in the sky will reach all television receivers. A small, inexpensive dish antenna aimed toward a location over the equator will be sufficient to receive the signal in the home. Such use of a communication satellite for direct broadcast to the home is called *direct broadcast satellite* (DBS) service. About six television channels could be broadcast in this way over one DBS satellite.

DBS service has thus far failed to become commercial in the United States. A number of reasons might explain this failure. There are already so many sources of television programs that one more source might simply be too many. DBS service by television networks would avoid the local affiliates, which are a source of local-interest programming. A DBS has all the image of a "big brother" in the sky. The DBS with its high-power transmitter is very costly.

SATELLITE MASTER ANTENNA TELEVISION (SMATV)

Large apartment complexes, hotels, and institutions are frequently wired with a private coaxial cable system to distribute television broadcasts received from a master antenna located on the roof. Such systems are called *master antenna television* (MATV) systems. Some newer MATV systems use a satellite dish to receive the television programs usually distributed to CATV operators. These systems are called *satellite master antenna television* (SMATV) systems.

SUBSCRIPTION TELEVISION (STV)

A conventional broadcast VHF or UHF television channel can be used to carry a pay-TV program. Such an application usually uses the relatively uncongested UHF spectrum in the evening and is called *subscription television* (STV). The STV subscriber pays a monthly fee to the STV operator and in return receives a device to descramble the broadcast signal.

The major problem with STV is that only a single channel is offered, as opposed to the much wider variety of programming available over CATV. Also, UHF reception is frequently not that good in many areas. It is reported that STV had as many as 1.4 million subscribers nationwide in 1982, but this number decreased to only 900,000 toward the end of 1983 (*see* Weinstein, p. 149).

LOW-POWER TELEVISION (LPTV)

The FCC allocates television channels to VHF and UHF broadcasters so that the frequency spectrum does not become overcrowded, resulting in interference. Usually, the FCC will ensure that about 150 miles must separate stations that use the same frequency channel to eliminate interference.

The amount of interference is not only distance dependent, it also depends on the power of the transmitter. The FCC, therefore, started accepting licenses in 1981 for *low-power television* (LPTV) stations operating in the VHF and UHF bands that would not interfere with the higher-power stations encountered in major metropolitan areas. LPTV would be particularly useful in rural areas not served by large stations. LPTV is restricted to about 10–100 watts in the VHF band and can cover from 10 to 20 miles at this low power. The UHF band for LPTV use is restricted to about 1,000 watts. As a basis for comparison, conventional broadcast television stations transmit at a peak power of from 500 to 50,000 watts. The low power needed for an LPTV station means that much less money is needed for transmitter equipment as compared with conventional television broadcasting.

MULTIPOINT DISTRIBUTION SERVICE (MDS)

Multipoint distribution service (MDS) television is a microwave system for distributing one or more television programs directly to the homes of subscribers. The service operates in microwave frequency bands at about 2.1 and 2.6 GHz. At these high frequencies, line-of-sight reception is required from the transmitting antenna to the home antenna. In addition to a special microwave antenna at the home, a decoder is required to translate the frequencies of the received signal from the microwave range to an unused VHF station for viewing on the home TV set. Frequently, the MDS signal is scrambled to prevent unauthorized use, and a descrambler would then be built into the decoder.

Early MDS television offered only a single channel. Newer *multichannel MDS* (MMDS) television can offer as many as eight television channels utilizing spectrum space previously allocated by the FCC for the exclusive use of *instructional fixed television service* (IFTS).

Clearly, MDS television does not offer the large variety of programming available with CATV. Since an extensive coaxial cable distribution system is not required, however, an MDS television service is considerably less expensive to initiate. In some markets, MDS television can be a viable competitor to CATV and can further increase the overall variety of program sources available to the consumer.

References

Cunningham, John E., *Cable Television (Second Edition),* Howard W. Sams (Indianapolis), 1987.

Stern, Joseph L., "Cable Television (CATV) Systems," in *Electronics Engineers' Handbook,* Donald G. Fink (Editor-in-Chief), McGraw-Hill (New York), 1975, pp. 21-58 to 21-82.

Weinstein, Stephen B., *Getting the Picture,* IEEE Press (New York), 1986.

12. Multichannel Sound

Broadcast FM radio has had stereophonic sound for a number of years. It therefore is natural that a similar feature should be offered for broadcast television, particularly now that larger screen television receivers are available. A number of techniques were developed by various companies for broadcasting stereo audio over television. In early 1984, the Federal Communications Commission authorized multichannel-sound broadcasting supporting a system recommended by the Broadcast Television Systems Committee (BTSC) of the Electronic Industries Association (EIA). The EIA recommendation is based on a system developed by Zenith Radio Corporation with noise reduction technology developed by DBX, Inc.

TV multichannel sound utilizes a multiplexing technique that is very similar to the amplitude-modulated audio subcarrier used for broadcast stereo FM radio. These multiplexing schemes combine a number of signals by using each individual signal to modulate individual subcarriers. All the modulated subcarriers are then added together, and this summed signal is used to frequency modulate the radio carrier for transmission over the airwaves.

The sum of the left (L) and right (R) stereo channels is formed as $L + R$. This $L + R$ signal is bandlimited from 50 Hz to 15 kHz and frequency modulates the audio radio-frequency (RF) carrier. This monophonic signal is received and demodulated by standard nonstereo television receivers and thus ensures compatibility. The $L + R$ signal is frequency shaped prior to transmission to emphasize the high frequencies. This preemphasis of the higher frequencies is performed by a simple filter with a time constant of 75 μs. Upon reception at the receiver, the preemphasis is removed along with any high-frequency noise that might have corrupted the signal.

The $L + R$ signal directly frequency modulates the audio radio-frequency (RF) signal that is broadcast over the air. The maximum peaks of the $L + R$ signal cause a frequency deviation of ± 25 kHz of the frequency-modulated audio RF signal.

A stereo difference signal is formed as $L - R$ and is bandlimited from 50 Hz to 15 kHz. The stereo difference signal amplitude modulates an audio subcarrier at twice the horizontal line frequency F_H of 15,734 Hz or 31,468 Hz. The audio subcarrier at 31,468 Hz is suppressed. The result is a double-sideband suppressed-carrier amplitude-modulated signal covering a frequency band of $31,468 \pm 15,000$ Hz, or from about 16.5 to 46.5 kHz, as shown in Figure 12-1. The maximum peaks of the stereo difference signal cause a maximum frequency deviation of ± 50 kHz of the frequency-modulated audio RF signal.

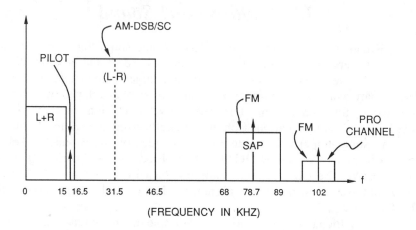

Fig. 12-1 Frequency spectrum of the various signals multiplexed together for multichannel television sound. All the various signals and modulated subcarriers are added together, and this signal then frequency modulates the audio radio-frequency (RF) carrier. The distance along the *x*-axis is proportional to the maximum frequency deviation of the RF carrier.

To reduce the effects of noise, the stereo difference signal is subjected to a variable encoding according to both its level and its frequency content. This type of level encoding for noise reduction was invented by DBX, Inc., and consists of wideband compression and spectral compression. The compression is applied at the transmitter, and the inverse of the process is performed at the receiver. Since compression at the transmitter is followed by expansion at the receiver, the overall process is called *compansion,* or companding. The wideband compression reduces the overall level of the signal equally for all frequencies. As the signal becomes larger, more compression is introduced, thereby reducing the dynamic range of the signal. The spectral compression reduces the high frequencies in the signal according to the high-frequency content of the signal. It is essential that the inverse expansion performed at the receiver be exact to minimize level and frequency distortion.

The television receiver needs to know whether a stereo signal is being transmitted. This is accomplished by transmitting a pure tone, or sine wave, at exactly one-half the frequency of the stereo subcarrier. This pure tone at 15,734 Hz is called the *stereo pilot signal.* Its maximum peaks cause a frequency deviation of ±5 kHz of the audio RF signal.

In addition to stereo audio, other signals can be placed in the audio RF signal through further multiplexing. Such services as bilingual sound, paging, and augmented audio for the blind are possible using a second subcarrier to achieve a *second audio program* (SAP). The subcarrier for the SAP is located at five times the horizontal scan frequency F_H, or 78,670 Hz. The SAP signal is bandlimited to 10 kHz, and its maximum peaks cause no more than a ± 10-kHz frequency deviation of the frequency-modulated SAP subcarrier signal. Double-sideband (DSB) frequency modulation of the subcarrier is used for the SAP. The SAP signal is subjected to compression similar to that used for the stereo-difference signal.

A third subcarrier can be used for narrowband audio or data signals. This third subchannel, called the professional subchannel or telemetry channel, utilizes a subcarrier at 6.5 times F_H or 102.3 kHz. The subchannel can be used for audio band-limited to 3.4 kHz or for data band-limited to 1.5 kHz. An audio signal is preemphasized using a filter with a 150-μs time constant, but no preemphasis is used for a data signal. The maximum peaks of the subchannel cause no more than a ± 3.5-kHz frequency deviation of the subcarrier frequency-modulated signal. Double-sideband frequency modulation (DSB) of the subcarrier is used for the professional channel.

The baseband $L + R$ signal, stereo pilot signal at 15,734 Hz, the subcarrier at $2F_H$ amplitude modulated by the stereo difference signal, the subcarrier at $5F_H$ frequency modulated by the SAP, and the subcarrier at $6.5F_H$ frequency modulated by the professional signal are all added together. This sum signal is then used to frequency modulate the audio radio-frequency (RF) carrier. The peak deviation of the audio RF carrier is ± 73 kHz.

References

Eilers, Carl G., "TV Multichannel Sound — The BTSC System," *IEEE Trans. Consum. Electron.*, Vol. CE-30, August 1984, pp. 236–241.

Hoffner, Randy, "Multichannel Television Sound Broadcasting in the United States," *J. Audio Eng. Soc.*, Vol. 35, September 1987, pp. 660–665.

Jurgen, Ronald K., "Stereophonic Sound For Television," *IEEE Spectrum*, September 1982, pp. 30–33.

Mitchell, Peter W., "Multichannel Sound," *High Technology*, April 1985, pp. 30–32.

Perry, Tekla S., "Consumer Electronics," *IEEE Spectrum*, January 1985, p. 77.

13. New Display Technology

INTRODUCTION

Futurists have predicted a large-screen television receiver for the home that would virtually envelope the viewer with visual stimuli. The large-screen would fill the wall of the viewing room in a fashion similar to large-screen movies at cinema theaters. Coupled with stereophonic sound and possibly even stereoscopic 3D imagery, television would be a multimedia event of considerable sensory appeal and involvement.

This chapter describes the technological advances that have been made in large-screen television, such as projection TV and large flat panels. Some technological approaches to 3D television are also described. The chapter ends with a description of some of the technologies for thin, flat displays of television images.

PROJECTION TV

There are two markets for projection television. One is the home, and the other is for industry, such as cinema theaters. Projection television for such industrial applications as the cinema theater has to produce a very large image with considerable brightness and good resolution. The requirements for projection television for the home are not so stringent.

There are three methods for projecting a television image onto a screen. The simplest method is to place a large lens in front of a television CRT, as shown in Figure 13-1. This method is called the *refractive technique*. It is difficult to achieve a bright image with good resolution with the refractive method, since the lens must be fairly large with a high speed. The refractive approach is used by Sony with a special Trinitron color CRT. The need for the CRT to create a very bright image results in considerable heat, and liquid cooling of the CRT faceplate and phosphor must be used in some systems.

The second method uses a *reflective optical system* to project the image on the faceplate of the CRT. As shown in Figure 13-2, the CRT image is reflected from a spherical mirror and then passes through a correcting lens to correct optical aberrations. This type of optical system is called a Schmidt optical system and is similar to the optical system used for large astronomical telescopes. It offers a high optical efficiency.

The electron gun, phosphor screen, and optical assembly can all be sealed together within one glass envelope, as shown in Figure 13-3. Three tubes, one for each primary color, would be used in a projection system. This sealed Schmidt system is used in a projection television system for

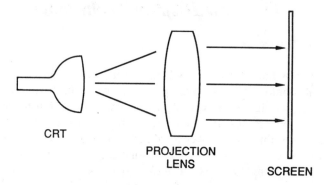

Fig. 13-1 Refractive technique for projection TV. The image on the face of the CRT is projected onto the screen through a large lens.

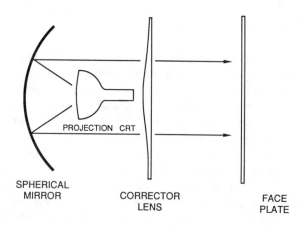

Fig. 13-2 Large-screen TV achieved by a reflective Schmidt optical system.

the home invented by Henry Kloss and introduced by the Advent Corporation in 1972. This *light-guide* tube produced a seven-foot diagonal image and opened the home market for large-screen projection television.

The third method is called a *light valve* and is used mostly in industrial applications. The basic idea of the light valve is to use an electron or laser beam to produce the television image on a material that then controls light from a high-intensity lamp or other source. The television image written on the material thus acts as a "valve" to control the light from the high-intensity source. The light valve might be transmissive with the light from the high-intensity source passing through it. Alternatively,

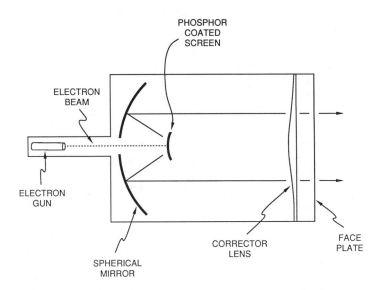

Fig. 13-3 In the Advent light-guide tube the elements of a reflective optical system and the CRT are all sealed together within a single tube.

the light valve might be reflective, in essence, a mirror that reflects light according to the television image written on it.

The first projection television system was the Eidophor, invented in 1939 by Professor Fischer in Switzerland. The Eidophor is a light-valve projection system. As shown in Figure 13-4, the television electron beam deforms a thin layer of oil on the surface of a concave mirror, and this results in a deformation pattern of the oil layer that corresponds to the television image. A high intensity light source is reflected off the surface of the oil-layer mirror, and the light is slightly deflected differently according to the deformation pattern of the oil layer. The light source is reflected onto the surface of the oil-layer mirror by a grid of mirror bars. The deformed reflection from the oil-layer mirror passes through the slits between the mirror bars and is projected onto the viewing screen. The light source in the Eidophor system typically is in the order of two kilowatts.

The projection screens used to view television images are specially constructed to produce a bright image. These screens are highly reflective and concentrate the reflected light to provide a light "gain," although obviously the total light reflected cannot exceed the incident light. This gain in reflected light is usually achieved at the expense of a narrow

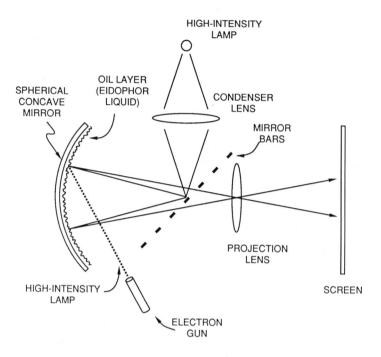

Fig. 13-4 Schematic of the Eidophor light-valve projection TV system. The electron beam writes the TV image by disturbing a thin layer of oil on the surface of a concave spherical mirror. A bright beam of light is reflected onto the mirror through a grid of mirror bars. The image is deflected by the disturbed layer of oil and passes through the slits between a grid of mirror bars. This image is projected through a lens onto the screen.

viewing angle. Some viewing screens are curved horizontally and vertically to improve the brightness and quality of the viewed image.

Rear projection is sometimes used with refractive projection television systems within a large cabinet to produce a large-screen television receiver for the home. The optical path is usually folded through the use of mirrors to make the cabinet more compact.

STEREOSCOPIC 3D TV

The fundamental principles of stereoscopic vision and three-dimensional (3D) displays apply to 3D TV. These principles are reviewed and are then applied to 3D TV.

Our two eyes view a scene from two slightly different directions. The result is that the brain receives two slightly different images of the same scene. The images are shifted differently in a horizontal direction, depending on the depth relationships between the various components of the scene. The human brain is able to translate these shifts into the sensation of depth, and we "see" the scene in three dimensions.

A camera can be used to take a picture of a scene from two slightly different directions or angles. If each picture is presented to each eye, the human brain will fuse on the scene and it will appear to have three-dimensional depth. Thus, the trick in creating a 3D display is the manner in which the two images are presented separately to each eye. One way of doing this is the old-fashioned stereoscope which uses lenses to enable each eye to focus on a card on which the two separate images are printed.

Another approach is to print the two images on top of each other, but in different colors, usually red and green. The viewer then views the two images through color spectacles. Although both images are shown to each eye, a red filter in one lens blocks the red image and only the green image is seen by that eye. A green filter over the other eye allows only the red image to be seen by that eye. Clearly, this approach, called an *anaglyph,* works only for black-and-white pictures.

A number of approaches have been investigated for showing moving images in three dimensions on a cathode ray tube. The screen can be split in half with the image for the left eye on the left side and the image for the right eye on the right side of the screen. A viewing hood is slipped over the CRT, and prisms are used to cause the eyes to diverge so that each eye sees only its corresponding image. This technique will work with black-and-white and color pictures. Rather than showing the two images on a split screen, two CRTs could be used with the image for the separate eyes shown on the separate CRTs. Miniature CRTs can be used with appropriate optics.

Another technique is based on the color anaglyph and uses a color CRT to display the image for one eye in red and the image for the other eye in green. Colored spectacles are then worn by the viewer. This technique works only for black-and-white pictures.

For television viewing, variants on the preceding methods are possible. Usually, the image for one eye is transmitted as one field, and the image for the other eye is transmitted as the other field. If one field is red and the other field is green, then the color-anaglyph approach will work. The viewer must wear spectacles.

Another approach requires special spectacles that perform as electronic shutters alternately opening and closing; while the image for the left eye is on the screen the left-eye shutter is open and the right-eye shutter is closed and *vice versa*. These electronic shutters are made of liquid crystal material and are controlled in synchrony with the frame shown on the television screen. Another approach is to place a liquid crystal polarizer over the screen that alternately polarizes each frame differently. The viewer need wear only polarized spectacles, with no need for electrical connection to the television receiver.

Two cameras are required at the studio for 3D television. The two images from these cameras must be multiplexed together somehow for transmission over the air. One way, previously mentioned above, is to send one image as one field and a second image as the other field of each frame. Unfortunately, this approach is not that compatible with existing non-3D receivers. An alternative approach would be to send one image over the assigned VHF channel and the other image over an available UHF channel. In this way, compatibility would be obtained.

FLAT DISPLAYS

The cathode ray tube was an essential ingredient in the success of television. Without it, a high-quality display of the moving image would not have been possible. The CRT, however, has a curved faceplate and is quite deep. As mentioned above, some futurists believe there is a market for large-screen displays that are thin enough to hang on the wall like a painting. Thin, small displays would be essential for small, portable, personal television receivers that could be carried in a pocket. Thus, the search for new, flat, thin display technology was begun.

The market for flat display technology falls into two application areas: television displays and computer displays. Displays used for television must be able to display moving images in color with good resolution. Computer displays are used mostly for alphanumeric text and stationary images.

The number of television receivers sold each year around the world is huge. Since each one of these television sets contains a cathode ray tube, the market for CRTs is likewise huge. In the United States alone, nearly $1 billion worth of CRTs were shipped in 1984. A replacement for the CRT for use in television receivers obviously would enjoy great commercial success. Thus far, however, the CRT continues to reign supreme over television display technology.

Some display technologies are active, or emissive, in the sense that they generate and emit light. Other display technologies are passive, or reflective, in the sense that they control light from some other source. The cathode ray tube is an example of an active display technology, since the phosphor screen emits light when excited by the electron beam. A liquid-crystal display is a passive display technology, since it can only reflect light from an external source.

There are a number of new display technologies that might be able to satisfy the need for thin, flat displays. Improvements and innovations in the cathode ray tube are being developed to make it both thin and flat. The liquid-crystal displays that are used in calculators are finding their way into small television sets. Such other technologies as ac and dc plasmas, electroluminescence, electrochromics, and electrophoretics seem most promising for the display of text for portable computers, but might some day find their way into television. The emphasis in this chapter is on display technology for television, and, hence, text-oriented display technologies are not described.

CONVENTIONAL CATHODE RAY TUBES (CRTs)

The cathode ray tube emits an electron beam from a heated cathode. This electron beam is focused and deflected across the rear of the viewing screen where it hits a phosphor coating, thereby causing light to be emitted. The elements of a CRT are in a vacuum enclosed in glass. The CRT has been around a long time and is more than one hundred years old.

A number of newer display technologies have been invented over the past few decades, and they have attempted to displace the old CRT. The CRT, however, continues to be the clear winner in nearly all application areas.

The CRT has a number of important advantages. It is easily manufactured, inexpensive, and reliable. It produces a bright, uniform image in color with a good gray scale. The CRT has a long life, in excess of 10,000 hours, and can produce a large image. The CRT, however, does have some disadvantages, too. It is large, bulky, and heavy with a fair amount of depth. It consumes a fair amount of power, generates heat, and requires high voltages. These disadvantages have led to the search for new display technologies to replace the CRT. With the exception of a few specialized markets, the CRT remains unchallenged.

The CRT is not a dead area for future technological development and innovation. The resolution of CRTs will continue to be improved. Fewer adjustments of convergence in color CRTs will be required. The color phosphors will continue to be improved. Larger screen CRTs will continue to be developed.

FLAT FOLDED-BEAM CATHODE RAY TUBES

Small, portable television receivers using a small, flat CRT were introduced a few years ago by the Sony Corporation and by Sinclair Radionics, Ltd., in the United Kingdom. The small, flat CRT used in these products had its germination in a flat CRT devised by William Ross Aiken in the early 1950s at the Kaiser Aircraft and Electronics Corporation in the United States. The Aiken CRT placed the electron beam at the side of the tube and caused it to bend toward the screen. The electron beam was folded, rather than following the straight path of the conventional CRT, and, hence, such tubes are called *folded-beam* tubes.

The flat CRT used in the Sony Watchman™ television receiver has the electron gun placed at the bottom of the tube, as shown in Figure 13-5. Electrostatic vertical deflection within the glass tube is used to cause the beam to deflect toward the phosphor screen. Horizontal deflection of the beam is by an electromagnetic yoke around the exterior of the neck of the tube. The phosphor screen is viewed through a transparent repelling electrode that forms the faceplate of the tube. A rectangular image is achieved through the use of electronic correction of shape distortions caused by the bent path of the electron beam entering from the bottom of the tube. Although this approach is practical for a small CRT, the flat faceplate would implode if a large-screen version were attempted.

A color CRT uses three beams, one for each of the three primary colors. A simpler method is to use a single beam and control it precisely as it passes over phosphor stripes of the three primary colors. This approach is called *beam indexing*. The position of the beam must be known precisely as it strikes each color phosphor. One way this is accomplished is by detecting the position of the beam with photodetectors built into the coating on the screen. Beam-indexed color CRTs have been built as prototypes for small-screen color television receivers, since the brightness needed for large-screen displays has thus far been a problem.

In the early 1950s, Dennis Gabor at the Imperial College in London investigated a flat, thin CRT in which the electron gun was placed behind the screen and pointed upward. This appears to be the germination of a flat CRT under development by Phillips Research Laboratories in England, and a prototype of the tube was described in 1982, although no product appears contemplated for the present. As shown in Figure 13-6, the electron gun is located behind the screen and is pointed upward. The low-energy electron beam is deflected horizontally by electrostatic deflection plates. The beam reaches the top of the tube and is then abruptly turned downward by the action of an electrostatic lens. Vertical deflection toward the screen is achieved by a large two-dimensional electrostatic

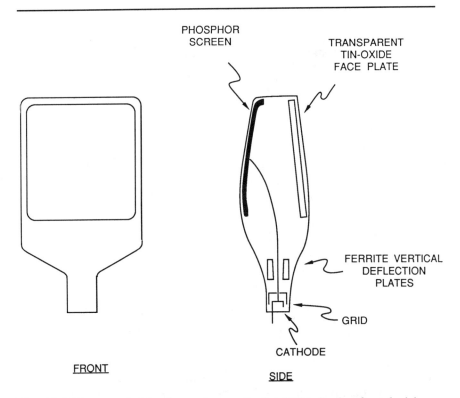

Fig. 13-5 Front and side views of a small, flat CRT display for television. The electron beam is at the bottom of the tube and is bent toward the phosphor screen by ferrite vertical deflection plates. Horizontal deflection of the beam is achieved by an external electromagnetic yoke. The image on the phosphor is viewed through the tube.

deflection plate. The electron beam is then amplified a thousandfold in beam current by an electron beam amplifier consisting of multiple electrodes in a two-dimensional array of horizontal channels. The amplified beam finally strikes the phosphor screen. With the exception of the glass faceplate, a metal enclosure is used for the tube, and the tube supposedly can be made large enough to produce a 20-inch diagonal picture.

CRT ADDRESSING

In some applications, it is necessary to be able to turn on and off a specific picture element, or *pixel,* in a CRT display. Such applications usually occur in the display of text and graphics.

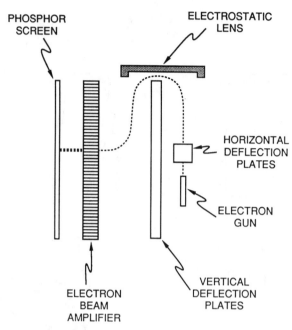

PHOSPHOR
SCREEN

ELECTROSTATIC
LENS

HORIZONTAL
DEFLECTION
PLATES

ELECTRON
GUN

ELECTRON
BEAM
AMPLIFIER

VERTICAL
DEFLECTION
PLATES

Fig. 13-6 A large, thin CRT is under development. A low-energy electron beam is placed at the rear of the tube. The beam is deflected and bent toward the front of the tube. Before striking the phosphor screen, the electron beam is amplified a thousandfold.

One approach is to apply the appropriate deflection voltages to the horizontal and vertical deflection yokes to cause the elctron beam to move to the desired pixel. The beam would then be moved to the next desired pixel, and so forth. If the beam were turned on while moving from one pixel to another, a straight line would be generated connecting the two pixels. This approach to addressing pixels is called the *directed-beam* or *vector* method. It produces a high-quality image. Time must be allowed to deflect the beam from one pixel to another, however, and since the display must be continuously refreshed at a fast enough rate to avoid flicker, the amount of information that can be displayed is limited.

Another method for addressing pixels is called the *raster-scan* method. With this method, the display is continuously scanned vertically and horizontally by the electron beam in a fashion identical to conventional television scanning. When the beam passes over a specified pixel, the beam is turned on for a small instant of time. The horizontal and vertical coordinates of the pixel must be converted into a particular instant of time for a particular horizontal scan line. This type of conversion is called *scan conversion* and is performed either by special hardware in the display or by software in a microprocessor in the display.

The display of textual information for such text-based services as teletext requires scan conversion of the text characters. A matrix of dots with a width of seven dots and a height of ten dots is allocated to each textual character. The actual characters are created using a 5 × 7 dot matrix; the extra horizontal and vertical dots are used for spacing and for descenders. Allowing for overscanning, about 215 scan lines per field can be used to display characters. This allows about 21 rows of text to be displayed. About 40 characters per row can be displayed. These 840 characters are stored in a page memory. The page memory is repetitively scan converted to control the raster display.

If the total area of the preceding text display is used for graphics, then it must be possible to access a single pixel. The total number of pixels in this example is 40 × 7 horizontally by 21 × 10 vertically, or 58,800 pixels. The information specifying individual pixels is stored in a "bit-map" memory, and this memory must be scan converted to control the raster display. If a gray scale is desired for each pixel, then the bit-map memory must be increased by the number of bits used to specify each level of gray, or color combination.

LIQUID-CRYSTAL DISPLAY TECHNOLOGY

Liquid crystal is a transparent crystalline material in a liquid form. The molecules in a crystal are all arranged in a symmetric manner. There are different liquid crystals according to the manner in which the molecules are arranged. One type that is frequently used for liquid crystal displays (LCDs) is the *twisted nematic type*. The term nematic refers to the elongated, cigarlike shape of the molecules. These molecules are all aligned in planes parallel to the surface of the liquid. The molecules at one surface are all twisted 90° compared to the molecules at the other surface. The change in the twist is gradual across the planes in the liquid. The molecules in a liquid crystal have the property that they will align themselves along an applied external electric field.

Light consists of waves with orientations in all planes. If light passes through a polarizer, only the light waves in one plane will be allowed to pass. The light is now polarized in some direction. If this light encounters another polarizer with its plane at 90° to the polarization of the polarized light, no light will pass through the second polarizer. The principle of polarization is used as the basis of operation of most LCDs.

One popular type of LCD consists of a twisted-nematic liquid crystal enclosed in glass, as shown in Figure 13-7. Transparent electrodes coat

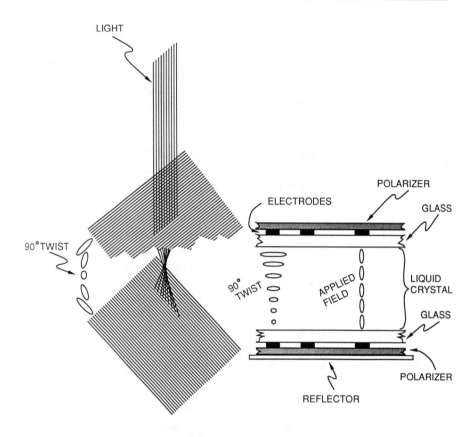

Fig. 13-7 A twisted-nematic liquid-crystal display contains crystals that twist and bend in the presence of an external electric field. As they twist and bend, light passing through them is bent 90°. An addressable display can be created through the use of polarizing filters.

the surface of the glass. Ambient light falls on the LCD and passes through a polarizer. As the polarized light passes through the twisted-nematic liquid crystal it is twisted 90° in polarization. The light then passes through another polarizer and is reflected back through the twist of the crystal to the viewer.

If an electric field is applied across the electrodes, the molecules all align themselves along the field. The twist is no longer present, and light passes through the crystal with no change in polarity. The light is now blocked by the rear polarizer, and no light is reflected back to the viewer. The display appears dark according to the pattern of the electrodes.

LCDs use considerably less power than CRTs. LCDs are thin and flat. They are well suited for small, portable television receivers. Compared to CRTs, however, they have lower contrast and resolution. This is because one portion of the liquid crystal can slightly affect adjacent portions. Fast-moving images can be blurred, since their response time is less than that for a CRT and a finite amount of time is required for the liquid crystal to change from one orientation to another.

A small liquid-crystal display is used in a portable black-and-white television receiver manufactured by the Toshiba Corporation. The display contains 220 × 240 pixels. A color liquid-crystal display for possible use in a small television receiver is being investigated by Toshiba. The display is back lit. The light passes through a polarizer, a twisted nematic liquid crystal, and another polarizer. Light that passes through the liquid crystal is filtered by a matrix of red, green, and blue color filters that form the color picture.

The challenge with using LCDs for television is the large number of very small pixels that are required to produce an image with good resolution. Another problem is maintaining the voltage at a particular pixel. This problem is solved through the use of a transistor and storage capacitor at each pixel. The transistor-capacitor combination is plated as a thin-film solid-state device on the glass surface along with the electrode.

LCDs have thus far been used for small portable television receivers. In theory, LCDs could also be used for large thin panels that would hang on the wall of a brightly lit room. Technological innovation, however, will be needed to make LCDs this large.

References

Aldersey-Williams, Hugh, "Flat Out for Pocket TV," *New Scientist,* May 5, 1983, pp. 282–285.

Apt, Charles M., "Perfecting the Picture," *IEEE Spectrum,* July 1985, pp. 60–66.

Baggett, James A., "How They're Shrinking Giant-Screen TV," *Popular Science,* November 1982, pp. 120–121.

Baldauf, David R., "The Workhorse CRT: New Life," *IEEE Spectrum,* July 1985, pp. 67–73.

Free, John, "3-D TV," *Popular Science,* June 1988, pp. 58–63, 110.

Kokado, Naoyuki, *et al.,* "A Pocketable Liquid-Crystal Television Receiver," *IEEE Trans. Consum. Electron.,* Vol. CE-27, August 1981, pp. 462–470.

Levine, Marty, "Theater-Size Home TV," *Popular Science*, February 1988, pp. 76–79.

Matsuzaki, Atsushi, *et al.*, "A Color Flat CRT and Its Application," *IEEE Trans. Consum. Electron.*, Vol. CE-32, August 1986, pp. 194–201.

Mitchell, Peter W., "Tiny TV's Hit It Big," *High Technology*, June 1984, pp. 20, 22.

Noll, A. Michael, "Real-Time Interactive Stereoscopy," *SID Journal*, Vol. 1, September-October 1972, pp. 14–22.

Ohkoshi, Akio, *et al.*, "A Compact Flat Cathode Ray Tube," *IEEE Trans. Consum. Electron.*, Vol. CE-28, August 1982, pp. 431–435.

Perry, Tekla S., "From Lab to Lap," *IEEE Spectrum*, July 1985, pp. 53–59.

Robertson, Angus, "Projection Television," *Wireless World*, Vol. 82, September 1976, pp. 47–52.

Robertson, Angus, "Projection Television — Refractive Projectors," *Wireless World*, Vol. 82, October 1976, pp. 67–72.

Smith, Kevin, "CRT Slims Down for Pocket and Projection TVs," *Electronics*, July 19, 1979, pp. 67–68.

Smith, Kevin, "Electrons Make U-Turn in Flat CRT," *Electronics*, October 20, 1982, pp. 81–82.

Tannas, Lawrence E., Jr., and Walter F. Goede, "Flat-Panel Displays: A Critique," *IEEE Spectrum*, July 1978, pp. 26–32.

Todd, L. T., Jr., and S. Sherr, "Projection Display Devices," *Proc. SID*, Vol. 27, 1986, pp. 261–268.

"CRT Displays Full-Color 3-D Images," *Information Display*, November 1985, pp. 12–13.

"Thin, Low-Power Display Rivals Standard CRTs," *High Technology*, Vol. 2, September-October 1982, pp. 92, 94.

14. CCD Image Sensors

PHOTODIODES

The basic element that responds to light energy in a solid-state imaging system is the *photodiode*. A photodiode generates an electric current that is proportional to the intensity of the light falling on it.

A photodiode consists of a junction formed between an *n*-type semiconducting material and a *p*-type semiconducting material, as shown in Figure 14-1. An *n*-type semiconductor has a majority of electrons — negative charges called *n*-type carriers — that support the flow of electric current. A *p*-type semiconductor has a majority of "holes" — missing electrons or positive charges called *p*-type carriers — that support the flow of electric current. Semiconductors are made from pure silicon to which small quantities of impurities are added in a process called *doping*. For example, the introduction of boron atoms into silicon will make it a *p*-type semiconductor.

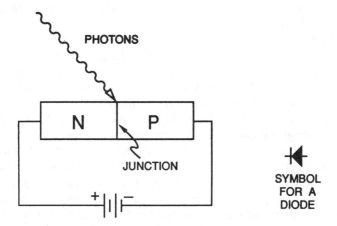

Fig. 14-1 A photodiode is formed by a junction of a *p*-type semiconductor and an *n*-type semiconductor. When exposed to light and a suitable external voltage, a photodiode generates an electric current proportional to the light intensity falling on it.

A semiconductor diode will conduct electric current in one direction only, and a diode conducting current is said to be *forward biased*. When an external voltage is applied in the opposite direction, the diode is *reverse biased* and does not conduct current. Electric charge is drawn away from the junction when a diode is reverse biased, and there thus are no majority carriers to conduct current across the junction.

When a semiconductor junction is exposed to light, extra hole-electron pairs are created. They separate and diffuse across the junction, thereby creating an electric current that increases with light intensity. A photodiode can also be visualized as a resistance that decreases with light intensity. A photodiode is usually operated in a reverse-biased fashion.

CCD IMAGE SENSORS

An imager can be constructed from a two-dimensional array of photodiodes. The image is focused on the array by conventional optics. The photodiodes respond to the light energy by creating hole-electron pairs in proportion to the light energy falling on each individual photodiode. The array in effect dissects the image into a number of very small picture elements, called pixels. The photodiodes average and store an electric charge that is proportional to the light energy falling on them. After an appropriate amount of time, the stored charge is read out from the photodiodes and creates an electric current corresponding to the image focused on the array.

The charge stored in each photodiode is read out through a transistor used as a switch that allows the electric charge stored in the photodiode to flow to an output conductor and thence to the final output circuit, as shown in Figure 14-2. Each photodiode has its own field-effect transistor (FET) used as a switch. An FET transistor has three terminals called the gate, the source, and the drain. When used as a switch, the application of a voltage at the gate will allow current to flow between the other two terminals. Otherwise, no current flows between the two terminals. The FET thus performs as a voltage-controlled switch. A voltage pulse is applied to a horizontal conductor, or line, and this voltage causes, or triggers, the transistor to conduct the charge stored in the photodiode. The charge flows through the transistor to a vertical output line.

In an actual image sensor, each photodiode and transistor forms an integral subelement, or pixel, in a two-dimensional array. The electric charge generated by the photodiode is stored in each pixel in potential "wells" in the circuit.

The array of photodiode and transistor pixels is organized into rows and columns, as shown in Figure 14-3. The gates of all the transistors in one row are connected to the same horizontal trigger line, and the outputs of all the transistors in one column are connected to the same vertical output line. Normally, all the transistors are in their nonconducting or "off" state. A vertical shift register is used to apply a trigger pulse to

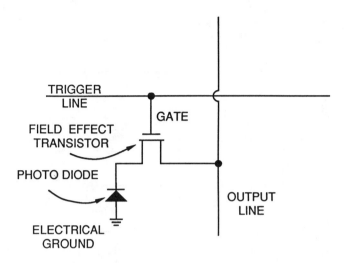

Fig. 14-2 A photodiode and a field-effect transistor form a light-sensing element that can be accessed by a voltage pulse applied to a trigger line. The electric charge created by the light falling on the photodiode is read out through the transistor which acts as an instantaneous switch in response to the trigger pulse.

one trigger line after the other in sequence from the top to the bottom of the array. The trigger pulse applied to a particular trigger line causes all the transistors in that row to connect each photodiode in the row to its corresponding vertical output line. The vertical output lines are connected to a horizontal shift register, which then shifts the electric charges from the photodiodes to the final video output line. The final video signal is smoothed by a low-pass filter to eliminate the small steps between the charge samples.

The horizontal shift register is a *charge-coupled device* (CCD) capable of shifting analog signals. The analog charge is sequentially passed along the device by the action of stepping voltages. The charge is stored as minority carriers in potential "wells" formed under electrodes. The charge is moved from one well to an adjacent well by the stepping voltages. The voltage applied to each electrode causes a potential well to form underneath it in a minority-carrier channel. A CCD is an analog shift register.

The trigger pulse that connects the photodiodes to the horizontal shift register occurs during the horizontal blanking interval. The electric charge stored in the horizontal shift register is read out in the time of the

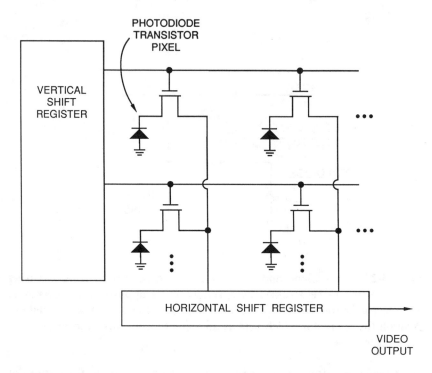

Fig. 14-3 An array of photodiodes and transistors is organized into rows and columns to create a charge-coupled device (CCD) image sensor. A vertical shift register sends a trigger pulse along each row in sequence. This causes the stored charge in each photodiode along the row to be transmitted along the vertical output line to a horizontal shift register. The information stored in the horizontal shift register is then shifted out and becomes one scan line of the video signal.

visible portion of a horizontal scan line and is the video waveform. Typically, every other row of photodiodes is accessed so that the odd-numbered rows form one field and the even-numbered rows form the other field. Each photodiode accumulates and stores electric charge for one frame before it is accessed and emptied to begin accumulating charge again.

A high-quality CCD image sensor has its pixels organized into 520 vertical columns and 483 horizontal rows. The total number of pixels in such an arrangement is over 250,000. All these pixels and associated shift registers are on a tiny chip less than one inch square.

COLOR CCD IMAGE SENSORS

Three CCD image sensors can be used to create a color camera through the use of mirrors or prisms and color filters to separate the image into its three primary colors. This approach is used in most professional color-television cameras. There is a way, however, to use a single CCD image sensor to obtain the three primary-color signals. This is accomplished by placing an elaborate tiny mosaic of microscopic color filters over the CCD image sensor.

One approach is a checkerboard pattern of red, green, and blue filter elements, as shown in Figure 14-4 (*see* Dillon, *et al.*). Because green alone can be used to estimate the luminance signal, green is used for half of the filter elements. The remaining half is evenly divided between red and blue.

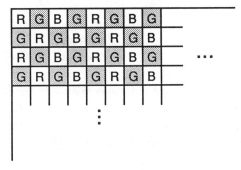

Fig. 14-4 One type of color CCD image sensor utilizes a checkerboard pattern of transparent red, green, and blue color filters. Each filter element covers a single photodiode in the CCD array. Since green alone can be used as a good estimate of the luminance signal, green is used for half the filter elements.

Another approach is a pattern of filter elements that pass all three primaries *(W = R + B + G)*, red and green *(Y = R + G)*, blue and green *(C = B + G)*, and green alone *(G)* (*see* Takemura and Ooi). This pattern is shown in Figure 14-5 for the scan lines of one field and would be repeated for the other field. The output for the odd-numbered scan lines is the sequence *W, G, W, G, . . . ,* and the output for the even-numbered scan lines is the sequence *Y, C, Y, C, ...* The luminance signal is thus obtained from alternate samples of the odd lines, since *W = R + B + G.* The green signal is obtained from the other alternate samples of the odd

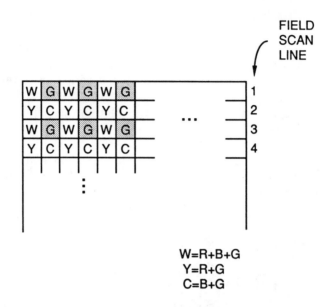

FIELD
SCAN
LINE

W=R+B+G
Y=R+G
C=B+G

Fig. 14-5 Another checkerboard pattern consists of filter elements that pass white *(W = R + B + G)*, red and green *(Y = R + G)*, blue and green *(C = B + G)*, and green alone *(G)*. The three primary signals are obtained by suitably delaying and subtracting alternate scan lines, as described in the text.

lines. Subtracting alternate samples of the even lines from the odd lines gives the blue signal, since $W - Y$ is equivalent to $(R + B + G) - (R + G)$ which equals B. Performing the same subtraction but delaying the even line by one pixel gives the red signal, since $W - C$ is equivalent to $(R + B + G) - (B + G)$, which equals R.

The color filter array must be aligned precisely above the appropriate pixels, which is clearly a tricky proposition, given the microscopic size of the filter elements and pixels. Because each filter element is so very small, light passing through one element can affect more than one pixel, thereby causing some blurring of the color resolution. Single CCD color image sensors are thus most suitable for very small television cameras for the consumer market.

SOLID-STATE TELEVISION CAMERAS

CCD image sensors offer a number of advantages over vidicon and other electron-tube imagers. They are light weight and very small in size, simple and rugged, require low voltages, and they are reliable, thereby eliminating the need for replacement. CCD image sensors offer high sensitivity, even in low light conditions, and burn-free operation in high light conditions. They are inexpensive. CCD image sensors are used in all television cameras for the consumer market.

Professional television cameras are currently starting to use solid-state CCD image sensors. A prism and color filters are used to separate the image into its three primary colors, and a high-resolution CCD image sensor is used for each primary color. Resolutions of 520 lines are possible along with high signal-to-noise ratios of over 55 dB. The image sensors use high-density very-large-scale integrated (VLSI) circuit technology with over 250,000 picture elements in a single sensor.

References

Childs, Ian, and J. Richard Sanders, "An Experimental Telecine Using a Line-Array CCD Sensor," *SMPTE Journal,* Vol. 87, April 1978, pp. 209–213.

Dillon, Peter L. P., David M. Lewis, and Frank G. Kaspar, "Color Imaging Using a Single CCD Area Array," *IEEE Trans. Electron Devices,* Vol. ED-25, February 1978, pp. 102–107.

McGinty, Gerald P., *Video Cameras: Operation and Servicing,* Howard W. Sams (Indianapolis), 1984, pp. 57–63.

Takemura, Yasuo, and Kazushinge Ooi, "New Frequency Interleaving CCD Color Television Camera," *IEEE Trans. Consum. Electron.,* Vol. CE-28, November 1982, pp. 618–623.

Weimer, Paul K., Michael G. Kovac, Frank V. Shallcross, and Winthrop S. Pike, "Self-Scanned Image Sensors Based on Charge Transfer by the Bucket-Brigade Method," *IEEE Trans. Electron Devices,* Vol. ED-18, November 1971, pp. 996–1003.

15. Uses of the Vertical Blanking Interval

INTRODUCTION

The vertical blanking, or retrace, interval is the interval of time that is allowed for the electron beam to retrace itself vertically from the bottom of the picture tube back up to the top. During this interval the beam is turned off, or blanked out. A total of 21 scan lines is allowed for the *vertical blanking interval* (VBI) in each field of NTSC television. In modern television receivers, the beam reaches the top of the picture tube in much less than the time interval of 21 scan lines. The first nine lines of the vertical blanking interval are used for vertical synchronization and equalization pulses. The remaining scan lines can thus be used to transmit other information before the visible portion of the picture begins.

VERTICAL INTERVAL TEST SIGNALS (VITS)

Television broadcasters give considerable attention to the technical quality of the signals they broadcast. Various technical tests of the signal quality are possible with appropriate test signals. Some of these test signals are placed in the vertical blanking interval, using lines 17 through 19 of each field. These signals are called *vertical interval test signals,* or VITS for short.

The VITS are used primarily to evaluate the overall performance of any transmission medium that might carry the television signal. The VITS can be used to evaluate such items as any distortion in the gain-frequency characteristics, luminance nonlinear distortion, phase distortion, and chrominance-to-luminance gain and delay inequalities. The VITS are inserted by the network so that the affiliated stations can evaluate the technical quality of the received signal prior to rebroadcast. This type of technical evaluation is particularly useful, now that television signals are so frequently transmitted over various transmission media such as satellites and terrestrial microwave. The VITS are removed by the local station before the television signal is rebroadcast.

The FCC has been tasked with formal regulation and control of the technical quality of the broadcast television signal and requires periodic technical tests and monitoring by the broadcasters. The records and logs maintained by the broadcasters are subject to review by the FCC. To save on costs, some VHF and UHF television stations operate remotely, and the FCC then requires the remote station to broadcast a specific set of VITS on lines 17 and 18.

VERTICAL INTERVAL REFERENCE (VIR) SIGNAL

A major problem with the NTSC color television system is that the color information is subject to the effects of delay, and this requires the use of hue and saturation controls at the television receiver. Modern television receivers utilize circuits that are able to monitor a special reference signal that is inserted on line 19 of both fields and are able to correct the displayed color automatically. This reference signal is called the *vertical interval reference signal,* or VIR signal for short.

The VIR signal contains such information as the color burst, sync, black reference, chrominance reference, and luminance reference. The television receiver locks onto the VIR signal and adjusts the color circuits accordingly to compensate for color variations from different program sources and stations. The consumer has no need to make adjustments using the hue and saturation controls. Most television stations have been broadcasting the VIR signal since the late 1970s, and most modern television receivers contain circuits that use the VIR signal. With the use of the VIR signal, a major disadvantage of the NTSC color television system has been eliminated.

CLOSED CAPTIONING

The video signal has a bandwidth of about 4 MHz. During the VBI, this bandwidth is mostly unused. This bandwidth could be used during this interval for digital data, which could be sent at a rate of about 6 million bits per second (Mb/s). A scan line can contain about 53 μs of useful information. Thus, if one scan line per vertical blanking interval is used for digital data, it can contain about 320 bits, or, equivalently, 40 characters of ASCII text. Since there are 60 blanking intervals per second, a single scan line per VBI could be used to transmit digital data at an average rate of 2,400 characters per second.

This, then, is the basic idea that led to the use of the vertical blanking interval to carry text. The first such use of the VBI was to carry a textual transliteration of the spoken dialogue of the audio channel. This textual information was then displayed as text superimposed along the bottom of the television picture for use by the hearing impaired. A simple adapter is available to decode this textual signal which is inserted in line 21 of one field and to display the text along with the conventional television picture. This use of the VBI for the hearing impaired is called *closed captioning* and was pioneered by the Public Broadcasting Service (PBS). The use of the VBI for closed captioning was authorized by the FCC in 1976.

The system used for closed captioning transmits only two eight-bit characters in a single scan line, although the preceding showed that as many as 40 characters could be sent per scan line. The reasons for the conservative use of technology with closed captioning included a desire to keep the design simple and the costs low so that the decoders would be inexpensive in small quantities. It was also important that the closed captioning signal in the VBI be capable of recording on video tape, and this dictated a fairly conservative approach.

TELETEXT

The basic idea of transmitting text during the vertical blanking interval can be used as a source of textual news and other text-based services that can be displayed on the home television receiver. These types of service are called *teletext* and were pioneered by the British.

The British Broadcasting Company (BBC) began experimental teletext transmission in March 1973, followed by experimental service in September 1974. The Independent Television (ITV) stations began experimental service of their teletext service in mid-1975. The BBC teletext service is called CEEFAX, and the ITV teletext service is called ORACLE. The British government gave final approval to teletext broadcast service in November 1976.

During the first few years of service from 1976 to 1979, there were fewer than 5,000 teletext-equipped television receivers in use (*see* Leith). In 1979, this increased to about 40,000 sets in use. Television set manufacturers then started to incorporate the teletext decoders within the sets, and, not surprisingly, the number of teletext sets in use greatly increased. In 1981, there were 250,000 sets in use. In 1982, there were more than 500,000 sets in use, and sales of teletext sets were about 8,000 sets per month.

The British teletext systems display a "page," consisting of 24 rows of 40 characters each, on the television screen. The 40 characters for each row along with an additional five bytes for headings and timing information are transmitted in one scan line. Lines 17 and 18 of each field are allocated to teletext. Since the field rate used by the British is 50 fields per second, a total of 100 rows of teletext are transmitted per second. A teletext page consists of 24 rows, and thus about four teletext pages are transmitted per second.

The pulses that specify each bit are sent at a rate of 6.9375 million bits per second on each scan line. A simple *nonreturn to zero* (NRZ) system is used to specify each bit. A binary zero is specified as the level of reference black, and a binary one is specified as 66% of the reference white level.

The teletext pages are transmitted in a round-robin fashion in which all the pages are transmitted in sequence over and over again, as depicted in Figure 15-1. If the total number of transmitted pages were 100 pages, for example, the 100 pages would be constantly repeated for transmission in a serial fashion. Since four pages are transmitted in one second, it would require 25 seconds to send the total 100 pages.

Each teletext page is identified with a page number. The user has a keypad and enters the number of the desired page. When that particular page is received, circuitry in the television receiver grabs it, stores it, and displays it on the screen of the television receiver. The storage buffer need only store the 960 ASCII characters that define a page. These characters are then converted into the dot patterns needed to turn the electron beam on and off as it scans across the TV screen. This process is called scan conversion and is usually performed by appropriate circuitry in the teletext decoder within the television receiver.

For the preceding example of 100 pages, it would take 12.5 seconds on the average to access a particular page. Information with a high demand could be repeated within the 100 frames to decrease the access time. For example, a high-demand page might be repeated every 20 pages, and the average time to access this page would be only 2.5 seconds. A total of 100 pages is called a *magazine,* and the total capacity of the British teletext system is eight magazines.

The type of information broadcast in teletext is mostly news, weather, sports, and financial information. In many ways, teletext can be envisioned as an on-demand textual form of radio. In the morning, we could use teletext to check the weather forecast and traffic information. In the evening, we could check sports and news. During the day, information about community and school events would be broadcast, and a television schedule might also be transmitted.

Textual characters are created on a display as a matrix of dots, as shown in Figure 15-2. The use of a dot matrix that is five units wide and seven units high is fairly standard. The horizontal space between characters adds one unit to the horizontal width, and the need to allow for descenders for some characters along with vertical spacing adds three units to the vertical height. Thus a total space of 6×10 must be allowed in the display for characters composed from a 5×7 dot matrix. A 5×7 character is contained within a 6×10 matrix.

British teletext offers seven colors: red, green, yellow, blue, magenta, cyan, and white. These colors can be used for the displayed character or for the background.

A crude, but very effective, form of graphics is used with the British teletext system. The 6×10 matrix is divided into six contiguous mosaic

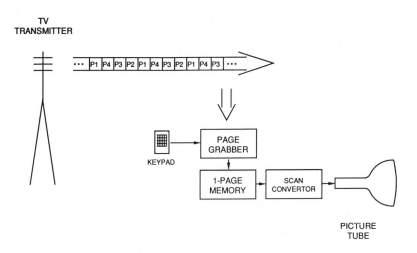

Fig. 15-1 The pages of teletext information are transmitted over and over again in a round-robin fashion. The teletext information is inserted in the vertical blanking interval of the broadcast television signal along with the TV signal. At the receiver, the user enters the number of the desired teletext page on a hand-held keypad. When the desired page comes by, it is grabbed and stored in a one-page memory. The characters that make up the page are then converted into the video signal needed by the television picture tube.

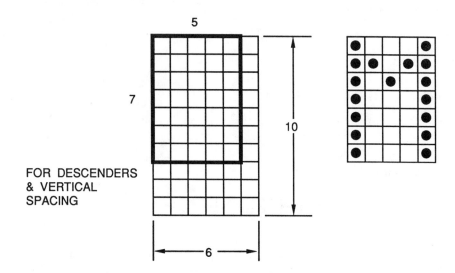

Fig. 15-2 Textual characters are formed from a 5 × 7 matrix of dots. The actual spacing is 6 units horizontally to allow for horizontal spacing and 10 units vertically to allow for descenders and vertical spacing.

blocks, as shown in Figure 15-3. Using an appropriate code, any one of the 64 various combinations of filled-in blocks can be specified. This type of graphics is called *mosaic graphics*. The blocks can be separated by one dot, and this is called a separated mosaic. A control character is used to specify either a contiguous or a separated mosaic.

Drawings created with mosaic graphics have a crude, blocklike appearance. An alternative approach is to define various graphic subunits, such as lines, circles, and rectangles, and to transmit the information needed to specify their size and location in the display. This latter approach is called *geometric graphics* and is based on computer graphics techniques. For example, suppose it is desired to draw a circle with a radius of R units centered at X units along the horizontal direction and Y units along the vertical direction. With a geometric graphics technique, the code for "draw a circle" is sent along with the specification of R, X, and Y. Electronic circuitry in the decoder then turns the electron beam on and off at the required locations to draw the circle as specified.

For a very simple image, geometric graphics is more efficient than mosaic graphics in transmission. The many geometric codes for a complex image, however, can consume more transmission than the fixed information required for mosaic graphics. It is time consuming to use mosaic

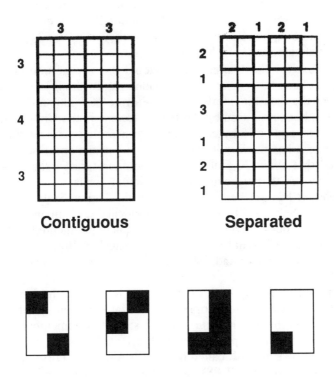

Fig. 15-3 The 6 × 10 matrix is divided into 6 blocks to form a mosaic. By filling in the blocks in different combinations, a total of 64 mosaics is possible. Four mosaics are shown. The blocks are touching in the contiguous form and are separated by one unit in the separated form.

graphics to create an image, and geometric techniques can be more efficient for many images. Clearly, it would, therefore, make sense to use the geometric technique to create the image, and this image could then be converted into a mosaic for transmission to the television receivers.

The size of the British teletext image must be changed for teletext in the United States. This is required because television in the United States consists of 525 scan lines of which only about 430 are actually visible on the screen after allowing for vertical blanking and overscanning. The number of visible lines per field is half that amount, or 215 lines. This means that 21 rows of text could be displayed on the television screen. European television consists of 625 scan lines of which about 540 are visible on the screen. This means that about 27 rows of text could be displayed, although actual European teletext systems use 24 rows of text to be on the safe side.

Teletext for the United States would use a page consisting of 21 rows of 40 characters each. Since the video bandwidth is less than British television, the data bit rate is less, and a rate of 5.2867 million bits per second has been proposed by the British for use in the United States. At this rate, 280 bits or 35 bytes would be transmitted per scan line.

Teletext has been quite successful in Europe and elsewhere. It is estimated that about 20% of TV households in Great Britain have television receivers that are equipped for teletext display.

Teletext has not been successful in the United States. The Public Broadcasting Service (PBS), National Broadcasting Company (NBC), and Columbia Broadcasting Service (CBS) have all experimented with teletext and offered experimental service in Los Angeles. Both NBC and PBS withdrew their service, and only the CBS ExtraVision® teletext service currently remains.

One reason for the lack of success of teletext in the United States involves standards. Some people believe that elaborate graphics are an essential requirement for teletext, but it should be remembered that the very simple mosaic graphics used in the British system did not hinder success. A new standard, called the North American Presentation Level Protocol Standard (NAPLPS), was proposed that offered an elaborate geometric-graphics capability based on computer graphics technology. The NAPLPS, however, requires longer transmission times along with more storage of the frames at the studio compared to the British system. The teletext decoders required for the television receivers are more complex and costly than the much simpler technology used in the United Kingdom. The confusion generated over standards and the need for sophisticated graphics certainly confused and stunted the development of teletext in the United States.

Television in the United States is advertiser-supported. The question then arises about how to sell teletext. Another concern is whether teletext viewing might distract from conventional television and its commercials. Clearly, one answer is to use teletext as a textual adjunct to commercials for information about sources of local supply for the product being advertised. In this way, a new source of revenue would be obtained for the affiliated stations. Teletext could also be used to promote programs that would be shown later in the viewing time, in essence, an electronic guide to television programs.

OTHER USES OF THE VBI

A 40-bit word is placed on line 20 in field 1 of the VBI as a network source code or identification. The month, day, hour, minute, second, frame, and program source are indicated within the code. The network source code is used by the broadcaster at the studio and is removed before broadcasting over the air.

The VBI is used to transmit text and graphics for SilentRadio service. The data are received by a special decoder and displayed as a single line of moving text and mosaic graphics. To maintain robustness and simplicity, a low data rate of 48 bits of information is sent on each VBI scan line. The service was first introduced in 1979 in the Los Angeles area and can use either one or two scan lines per frame. General-interest information is transmitted, and the displays can be seen mostly in restaurants, bars, and banks.

References

Hurly, Paul, Mathias Laucht, and Denis Hylnka, *The Videotex and Teletext Handbook,* Harper and Row (New York), 1985.

O'Connor, Robert A., "Current Usage of Vertical Test Signals in Television Boadcasting," *IEEE Trans. Consum. Electron.,* Vol. CE-22, August 1976, pp. 220–228.

Prentis, Stan, *Color Television Course,* Tab Books (Blue Ridge Summit, PA), 1972, pp. 88–95.

Tydeman, J., *et al., Teletext and Videotex in the United States,* McGraw-Hill (New York), 1982.

Veith, Richard H., *Television's Teletext,* Elsevier Science Publishing (New York), 1983.

16. High-Definition Television

Conventional television in the United States uses 525 horizontal scan lines per frame of which 42 are reserved for the vertical retrace interval. This leaves about 483 usable scan lines with a vertical resolution of about 338 lines taking into account the effects of interlacing on apparent resolution. The aspect ratio is 4:3, and the field rate of NTSC color television is very nearly 60 fields per second. To match the vertical resolution, the bandwidth of the luminance signal is 4.2 MHz. With these standards, the quality of the television image is approximately equivalent to super-8 motion picture film. In an attempt to improve this performance by making it closer to 35-mm film, a higher resolution television, called high-definition television (HDTV), has been proposed and is under active investigation.

There are a number of different systems and standards that have been proposed for high-definition television. Many of the differences stem from the different number of horizontal scan lines used by the television systems of different countries. Most HDTV proposals double the number of horizontal scan lines. For the United States and the NTSC color television standard, this results in 1050 scan lines. The field rate of 60 fields per second is unchanged, although the aspect ratio is increased in the horizontal dimension from 4:3 to 5:3.

To carry this increased amount of information, the bandwidth of the television signal must be increased nearly sixfold from its present 4.2 MHz to nearly 30 MHz. Clearly, this considerable amount of bandwidth makes HDTV difficult to achieve commercially either by over-the-air broadcast television or conventional video cassette recording.

One HDTV system is under investigation by the Japan Broadcasting Corporation (NHK). The NHK HDTV system requires a bandwidth of nearly 30 MHz. NHK has developed a bandwidth-reduced version that requires only 8 MHz. The bandwidth reduction utilizes a technique called *multiple sub-Nyquist sampling encoding* (MUSE). The full-bandwidth HDTV signal is sampled with a compression of four using a special two-dimensional sampling pattern. At the receiver, a frame buffer is used to store the various samples as they arrive over the time of four fields to reconstruct the high-definition image. With the MUSE compression, stationary images are given full resolution, while moving portions of an image are given only one-quarter resolution. This works well for all but fast-moving scenes, which would be somewhat blurred anyway. The image processing requires elaborate circuitry both at the television studio and also within the television receiver. The technique is not compatible with conventional television signals.

An alternative approach was suggested in 1983 by CBS Laboratories. Although this approach is no longer being pursued, its basic principles were quite simple and it suggested a way to achieve compatibility. The CBS HDTV system achieved compatibility with conventional television. This was accomplished by sending a conventional television signal over one channel and the additional information required to construct a high-definition image over a second separate channel. The CBS HDTV system had less resolution at the edges of the picture, but offered full 1050-line resolution throughout most of the picture.

RCA Laboratories, now the David Sarnoff Research Center, has demonstrated an *extended definition television* (EDTV) system that is compatible with conventional NTSC color television and does not require any additional bandwidth. The system, called *advanced compatible television* (ACTV), offers wide screen viewing with a 3 × 5 aspect ratio and 1050 lines. The extra information required for the wider aspect ratio is transmitted in two components. One component is pasted in 1-μs intervals on both sides of the video picture in the portion that would be normally overscanned and not visible on conventional television receivers. The second component along with additional horizontal detail is used to quadrature amplitude modulate a 3.1-MHz subcarrier that is frequency interleaved between the harmonics of the luminance signal. The information encoded in this subcarrier is shifted in phase in such a way that it is separable from the chrominance signal and is not visible on an existing NTSC receiver.

The extra information needed for increased vertical detail is transmitted in the form of a "helper" signal that is used to quadrature amplitude modulate the broadcast RF signal. Adjacent 525-scan lines are averaged, and the helper signal is then added to each averaged line to create the required extra lines for a 1050-line display. The system offers nearly a 40% improvement in resolution over conventional NTSC television and is claimed to be nearly as good as a true 1050-line system. If additional spectrum space is available, perhaps through the use of a second channel, additional information can be transmitted to increase the picture resolution.

There is a considerable amount of redundancy in a television signal. Most single images do not vary that much from one small portion to adjacent portions. Also, very little of the image varies from one frame to the next. These forms of redundancy can be utilized to reduce the bandwidth required to transmit the television signal. This is usually accomplished by converting the television signal to a digital format and by then using digital processing to remove much of the redundancy, thereby achieving a much lower bit rate.

An analog video signal is usually sampled at a rate four times the color subcarrier frequency or 14.32 million samples per second. The amplitude variation of the sampled video signal is quantized into 256 levels, which are then encoded by using eight bits. The final bit rate for the digitized video signal thus is $8 \times 14.2 M = 113.6$ million bits per second (Mb/s). This digital signal is processed by using a coder to remove much of the redundancy inherent in the television signal. The lower bit rate signal can then be transmitted to the receiver where a decoder can recreate the television signal. The equipment that performs the coding-decoding is called a *codec*.

A study using a variable bit rate codec with ten different entertainment video programs showed that "VCR quality" required a bit rate of no more than 1.2 Mb/s (*see* Judice and Jaquez). The 1.2 Mb/s digital signal could easily be transmitted over a broadcast television channel. Unfortunately, "VCR quality" is much less than standard NTSC television and is, indeed, quite poor. Although digital decoders, however, are presently too complex and costly for use in television receivers, someday it might be possible to transmit an encoded digital signal along with the extra information needed for HDTV within a reasonable amount of bandwidth.

High-definition television (HDTV) is only one form of improvement being considered for television. HDTV offers twice the number of vertical lines of conventional television along with an increased aspect ratio. As explained previously, HDTV requires a considerable increase in bandwidth which creates compatibility problems with conventional NTSC broadcast television.

An alternative approach is to extend the definition of the received television signal while maintaining compatibility with conventional NTSC broadcast television. These forms of extended-definition television usually offer an increase in the aspect ratio, but do not offer as much improvement in resolution as is obtained with full-bandwidth HDTV. The RCA system is considered to be an EDTV system.

Yet another approach to improving the image quality of television is simply to improve the circuitry and the picture tube of a conventional NTSC television receiver. This approach, called *improved television,* is clearly compatible with conventional NTSC broadcast television. One improvement is the use of a delay-line filter for the chrominance signal so that the full 1.5 MHz of the *I* signal can be utilized. Another improvement, being investigated by the Sony Corporation, is the use of a 1050-line picture tube with line averaging between adjacent 525-scan lines to obtain the additional lines. Although resolution is technically not increased, there is a definite improvement in the perceived quality of the image with line averaging, since most of the visible scan-line structure has thus been eliminated.

HDTV involves not only the television receiver. Improved cameras and video recorders will be needed at the studio. The present quality of the program material broadcast over NTSC television can sometimes be so poor that an improvement in this area alone would be quite worthwhile without the need for 1050-line HDTV. Also, 1050-line HDTV might not be sufficient improvement in the long run, particularly for very large screen television, and HDTV with 2000 or more lines might ultimately be needed for commercial success. Another possible problem with large screens is that 60 fields per second can result in visible flicker, and hence the field rate might need to be increased to 80 fields per second. Before anything significant occurs with HDTV, however, the standards issue must be addressed.

There is a tremendous installed base of television receivers throughout the world. Many consumers would probably purchase television receivers with improved definition and quality, particularly for large-screen receivers. The existing installed base, however, cannot be disenfranchised in the process through the forced introduction of a new, noncompatible standard. The initial experience with the CBS noncompatible color television system should teach the industry a lesson about the need for compatibility and a well planned strategy.

The whole area of HDTV has generated considerable excitement and controversy. Many of the proposed HDTV systems require more bandwidth than the 6 MHz assigned for conventional broadcast television. This might mean the slow decay and death of conventional broadcast television, since it is believed that most consumers would watch the HDTV programs available from other sources. On the other hand, program content might be far more important than image quality to many viewers, as indicated by the fact that old black-and-white Jackie Gleason shows are still quite popular.

It has been suggested that the FCC reserve additional spectrum space for the broadcasting of HDTV signals. Spectrum space is a precious commodity, however, and many other people want the spectrum space for services such as mobile telephony. Also, the UHF band is mostly unused for a number of possible reasons such as a lack of program material and also technical taboos against too much spectrum crowding. In the future, improved technology could make more of the UHF band available for such novel television uses as HDTV. For example, one HDTV scheme proposes to transmit the additional information required for HDTV reception over a UHF channel to augment the standard TV signal transmitted over a VHF channel.

Clearly, there are a considerable number of unresolved issues, but the most uncertainty involves the consumer, the television viewer. Will the viewer notice the improvement possible with HDTV, particularly when compared with other methods for improving the quality of reception of conventional broadcast television? How much will the viewer be willing to pay to view the HDTV signal?

HDTV has been promoted in terms of its ability to bring the quality of 35-mm movies to the home, along with wide screen viewing, but will consumers be willing to rearrange their living rooms to accommodate a five-foot wide television screen? Is television viewing a family or group affair, or has television viewing become more individualized with each viewer having a personal television set. If the latter, then a small-screen receiver that can be held in the lap might be more desirable than a large screen on the wall.

Small, personal television receivers are currently very popular. The personal television set is viewed at a close distance and thus clearly requires very high definition along with a thin, flat screen. Such a small-screen receiver would require high definition, since it is not the absolute size of the screen that is significant, but, rather, the height of the screen relative to the viewing distance. If a small HDTV videocassette recorder were included, then the broadcast HDTV standard issue could be circumvented. The receiver could interpolate lines when viewing conventional television to give an improved image.

What all of this implies is that enough uncertainties abound to mean that HDTV is still many years away from achieving significant market penetrations. HDTV, clearly, is the television of the next century, although, in fairness, it should be mentioned that many of its advocates believe it will be the TV of the 1990s. One must be prepared for the future, and thus the use of higher definition equipment in the studio to produce and record television programs would seem to make good sense.

References

Fujio, Takashi, "High Definition Television Systems: Desirable Standards, Signal Forms, and Transmission Systems," *IEEE Trans. Commun.,* Vol. COM-29, December 1981, pp. 137–146.

Gaggioni, Hugo, "The Evolution of Video Technologies," *IEEE Communications Magazine,* Vol. 25, November 1981, pp. 20–36.

Isnardi, M. A., *et al.,* "A Single Channel, NTSC Compatible Widescreen EDTV System," David Sarnoff Research Center (Princeton, NJ), 1987.

Judice, C. N., and M. J. Jaquez, "High Compression Coding of

Entertainment Video for Packet Networks," *Bell Communications Research*, 1987.

Jurgen, Ronald K., "The Problems and Promises of High-Definition Television," *IEEE Spectrum*, December 1983, pp. 46–51.

Jurgen, Ronald K., "High-Definition Television Update," *IEEE Spectrum*, April 1988, pp. 56–62.

Lechner, Bernard J., "Higher Resolution, Fewer Artifacts, TV Technology Goals," *Information Display*, December 1985, pp. 12–15.

Lu, Cary, "High Definition TV Comes At High Cost," *High Technology*, July 1983, pp. 45–48.

Lu, Cary, "High-Definition Television," *High Technology*, April 1985, pp. 32–36.

Schubin, Mark, "Television Redefined," *Technology Illustrated*, July 1983, pp. 32–35.

17. International Standards

INTRODUCTION

The NTSC system, the first commercially successful system for broadcasting color television, was adopted in 1953 by the Federal Communications Commission as the national standard for the United States. Since then, about twenty-five other countries, including Canada, Mexico, Japan, and Taiwan, have adopted the NTSC system.

From its invention, there were inherent problems with the NTSC system, primarily since the hue and saturation of the color were encoded as phase and amplitude modulation of the color subcarrier. This led to a sensitivity to phase and amplitude errors, thereby requiring the use of controls at the receiver to adjust the hue and saturation. The problem was so bad that NTSC came to be known jokingly as standing for "Never The Same Color."

Prior to the invention of color television, monochrome television in the United States was based on a field rate of 60 fields per second and a vertical resolution of 525 lines per frame. The field rate was based on the 60-Hz frequency of electric power. Other countries around the world that utilized 60-Hz electric power likewise adopted the 525 lines per frame as used in the United States. European countries used a 50-Hz electric power system, and they compensated for this lower field rate by using a higher vertical resolution of 625 scan lines per frame. The world thus divided itself into two television standards: 60 Hz with 525 lines and 50 Hz with 625 lines. The 60-Hz standard offered increased resistance to flicker, but less resolution. The 50-Hz standard offered increased resolution, but less resistance to flicker.

With the invention and availability of compatible color television in the United States, the Europeans and others wanted color television for themselves, too. Some countries, however, saw this as an opportunity to avoid the problems encountered with the NTSC system, and they, therefore, pursued newer approaches. Nationalism entered the picture, too, with the result that some countries pursued their own system just to be different. Two new standards finally emerged. The first, SECAM, was pioneered by the French. The other, PAL, was pioneered by the Germans. The nationalistic aspect was such an influence in the development of the new standards that SECAM was claimed to stand for "Something Essentially Contrary to the American Method."

The SECAM standard was initially adopted by France and also by the USSR. The PAL standard was initially adopted by the Federal Republic of Germany and also by the United Kingdom. The world, thus, has three different techniques for color television: NTSC, PAL, and SE-CAM. All three techniques are similar in the way they treat the luminance information and in that compatibility with monochrome receivers is achieved. The three systems are different in the way the chrominance information is encoded and transmitted. The technical details of the PAL and SECAM systems are described in this chapter.

THE PAL SYSTEM

The *phase alteration line,* or PAL, system for color television was developed by Doctor Walter Bruch of the Telefunken Company in the Federal Republic of Germany. Broadcast color television using the PAL system began in 1967 in the FRG and the United Kingdom. A total of about thirty-six countries, including Norway, Sweden, Spain, and Switzerland, have since then adopted the PAL system.

A major problem with the NTSC system is that the hue of the color is very sensitive to errors in the phase of the color subcarrier. This requires the use of a hue control at the television receiver to correct for such errors. The PAL system attempts to eliminate this problem by transmitting the color difference signals as quadrature modulation of the color subcarrier like the NTSC system, but by reversing the phase of the $(R - Y)$ signal on alternate scan lines. In this way, any phase errors cancel, thereby giving a more accurate rendering of the color transmitted. A PAL color television receiver does not need a hue control and has only a saturation control to adjust the received color.

A PAL color television receiver uses a delay line to store one scan line so that the averaging of the color signals can be performed electrically. During the early days of color television, this delay line was costly, but nowadays it is reasonably inexpensive.

The color difference signals $(B - Y)$ and $(R - Y)$ are each bandlimited to 1.3 MHz. The color subcarrier has a frequency of 4.433618 MHz.

The television systems used by the countries that have adopted the PAL color television system operate at 50 fields per second with 625 lines per frame. The horizontal scanning frequency is 15,625 Hz. For the PAL system used in the United Kingdom, the baseband video signal occupies a bandwidth of 5.5 MHz, and the audio signal is transmitted on a separate frequency modulated carrier located at 6.0 MHz above the video carrier. Vestigial-sideband amplitude modulation (AM) is used to broadcast the video signal. A broadcast television channel in the United Kingdom occupies a channel width of 8.0 MHz. The PAL system used for continental Europe has a video bandwidth of 5.0 MHz, and the audio carrier is 5.5 MHz above the video carrier.

THE SECAM SYSTEM

The color television standard used in France and the USSR is called *séquentiel couleur avec mémoire,* or SECAM for short, and translates as "sequential color with memory." The system was developed in the early 1960s in France by Henri de France of the Compagnie Française de Télé-vision, and broadcast service began there in 1967. A total of about twenty-two countries, including the German Democratic Republic, Poland, Egypt, and Saudi Arabia, have adopted the SECAM system since then.

The television systems used by the countries that have adopted the SECAM color television system operate at 50 fields per second with 625 lines per frame. The horizontal scanning frequency is 15,625 Hz. The baseband video signal occupies a bandwidth of 6.0 MHz, and the audio signal is transmitted on a separate amplitude modulated carrier located at 6.5 MHz above the video carrier. A broadcast television channel occupies a total channel width of 8.0 MHz, and vestigial-sideband amplitude modulation is used to broadcast the video signal.

Like the NTSC color system, the SECAM system utilizes color difference signals $(R - Y)$ and $(B - Y)$ which are transmitted along with the 6.0-MHz luminance signal. Each color difference signal is bandlimited to 1.0 MHz. The color difference information is encoded by frequency modulation of a color subcarrier using a fairly large deviation of the color-subcarrier frequency. The use of frequency modulation to encode the color information avoids the phase and amplitude distortions inherent in the NTSC system. Thus, a SECAM television receiver needs neither a hue nor a saturation control.

With SECAM, the $(R - Y)$ and $(B - Y)$ signals are transmitted sequentially as separate signals on alternate scan lines. The $(R - Y)$ signals are transmitted on odd lines, and the $(B - Y)$ signals are transmitted on even lines. A one-line memory is used to store one of the signals until the other is received so that they can then be combined to obtain the three color signals.

The color difference signals are used to frequency modulate the color subcarrier. Unlike the NTSC color system in which the color subcarrier is suppressed, the color subcarrier is always present with the FM system used in the SECAM system. A color subcarrier frequency of 4.25000 MHz with a maximum frequency deviation of 280 kHz is used for the $(R - Y)$ signal. A color subcarrier frequency of 4.40625 MHz with a maximum frequency deviation of 230 kHz is used for the $(B - Y)$ signal.

The use of different frequencies on each succeeding scan line minimizes the visibility of the signal on monochrome receivers. Monochrome compatibility is further assisted by reversing the phase of the color sub-carrier on every third scan line and also between each frame. The amplitude of the color subcarrier is reduced for low luminance level color values to reduce further the visibility of the color subcarrier.

POLARITY OF VIDEO MODULATION

The composite color signal broadcast in the NTSC system is reversed in polarity so that an increasing positive voltage corresponds to increasing levels of blackness. This is called *negative video modulation*. It is used to minimize the effects of noise, as well as for other reasons discussed in a previous chapter.

The SECAM system used in France uses *positive video modulation*, so that increasing excursions of the video carrier represent increasing levels of luminance. The SECAM system used in the USSR, however, uses negative video modulation. The PAL system is broadcast using negative video modulation in all countries.

References

Bruch, W., "The PAL Colour Television System," Telefunken A. G., Grundlagenlabor Hannover.

Compagnie Française de Télévision, *SECAM Colour T.V. System.*

Pritchard, D. H., and J. J. Gibson, "Worldwide Color Television Standards — Similarities and Differences," *J. Soc. Motion Picture and Television Engineering,* Vol. 89, February 1980, pp. 111–120.

Roizen, Joe, "The Status of International Television," *Video Systems,* October 1987, pp. 26–31.

18. The Future of Television Technology

Television in its many forms is the prime source of entertainment for most people in the industrialized world. There is no reason to expect this to change in the future.

Television is a system involving two aspects: distribution technology and receiver technology. Each technology is independent of the other, but the two are intertwined, since one cannot work without the other.

Currently, there are many media over which television programs are distributed. Television programs are available over the air *via* very-high frequency (VHF) and ultra-high frequency (UHF) transmission. Television programs are distributed to many homes over a coaxial cable operated by CATV companies. Movies can be rented from neighborhood stores for viewing on the home VCR machine. Large dish-shaped antennas are used to receive television programs broadcast over satellites. Microwave radio is used to transmit television programs to homes.

What is most interesting about the many forms of distribution is that the older forms simply make room to accommodate the newer forms. A newer form does not replace an older form. It would appear that the public's appetite for video entertainment from a wide variety of media is, indeed, insatiable.

One new medium that might change the replacement doctrine is optical fiber. Optical fiber direct to the home would offer considerably greater capacities than are available from any other broadcast medium. Optical fiber would clearly replace coaxial cable in CATV systems. The large capacity of the fiber might be a solution to the increased capacities needed for high-definition television. Since the same fiber can also carry telephone conversations, a reconfiguring of the separation and roles of the CATV operator and the phone company might occur with all sorts of regulatory and policy issues.

Receiver technology is equally exciting. A television receiver can be visualized as mostly a display device with some electronic circuitry. Transistorized integrated circuits have reduced the size and cost of the circuitry to almost nothing. What remains is the display technology. The cathode ray tube has been the technology of choice since the earliest days of television. Newer technologies such as liquid crystals, however, might challenge the CRT, although innovations in CRT design might very well assure its future for a few more decades. High-resolution displays will be required for high-definition television, and only CRT technology will be able to accomplish the needed resolution.

Television circuitry continues to improve so that more of the quality of the broadcast signal can be utilized at the receiver. Digital circuitry and high-definition television offer the promise of further improvements in the quality of the displayed image at the receiver. What this means is that there will be increased pressure to improve the quality of the transmitted image through the use of higher-quality cameras and program sources.

Television is an analog technology. Its invention, along with compatible color television, epitomized analog technology at its pinnacle. The world, however, now appears to be going digital. Optical fiber is essentially a digital transmission medium in which a beam of light is turned on and off to convey information. The compact disc has revolutionized the world of recorded sound. Will television, too, go digital in the future?

The television signal requires a bandwidth of 4.5 MHz, and straight conversion of this analog signal into a digital signal would require an extremely large bit rate. The television signal, however, does not change that much from frame to frame or from line to line. The considerable amount of redundancy in the television signal means that the bit rate can be greatly reduced through suitable processing at the transmitter and receiver. Today, such processing would be quite complex and costly. Processing power through microprocessors, however, continues to become less and less costly, so that one day digital transmission and recording of television signals might become commercially feasible. The result would be an improvement in the visual quality of the imagery.

The large, flat display that offers a wall of imagery in three dimensions is still far in the future. For the time being, we will have to be content with digital circuitry, improved phosphors, and high definition, all of which are closer, or already occurring.

As for holographic television, that is still farther away. Currently unforeseen new forms of television, however, might be invented along the way. For example, it would be nice, as a viewer, to be able to choose your receiver from among the different cameras used to televise an event. Someday this might be possible, so that the viewer, rather than the director, will decide which camera to view.

The future of television technology thus will continue to be most exciting with novel services and technological innovation. Television clearly is an evolving technology, and the course of its future evolution will be fascinating to witness. The demystification of television technology that this book seeks to accomplish for the reader will make such future evolution more understandable.

Bibliography

Grob, Bernard, **Basic Television and Video Systems,** McGraw-Hill (New York), 1984.

Howard W. Sams Editorial Staff, **Color-TV Training Manual,** Howard W. Sams (Indianapolis), 1980.

Ingram, Dave, **Video Electronics Technology,** Tab Books (Blue Ridge Summit, PA), 1983.

Kybett, Harry, **Video Tape Recorders,** Howard W. Sams (Indianapolis), 1978.

McGinty, Gerald P., **Video Cameras: Theory and Servicing,** Howard W. Sams (Indianapolis), 1984.

McGinty, Gerald P., **Videocassette Recorders: Theory and Servicing,** McGraw-Hill (New York), 1979.

Prentis, Stan, **Basic Color Television Course,** Tab Books (Blue Ridge Summit, PA), 1972.

Schure, Alexander, **Basic Television Volume 6,** Hayden Book Company (Rochelle Park, NJ), 1985.

Tinnell, Richard W., **Television Symptom Diagnosis,** Howard W. Sams (Indianapolis), 1977.

Glossary

This glossary defines terms and concepts that are not defined as they are used in the text.

ac — the abbreviation for *alternating current*. Alternating current is an electrical current that varies in amplitude and direction, or **polarity.**

amplifier — an electronic device that can make a signal larger in amplitude. The gain of an amplifier is usually expressed in *decibels,* or *dB*.

amplitude modulation (AM) — a technique for varying the maximum amplitude of a high-frequency tone, or *sine wave,* called the carrier, in synchrony with the instantaneous amplitude variation of an information-bearing signal. The effect of amplitude modulation is to translate the frequency spectrum of the information signal to a range of frequencies around the carrier frequency. The translated spectrum below the carrier frequency is called the **lower sideband,** and the translated spectrum above the carrier frequency is called the **upper sideband.** The lower and upper sidebands are mirror images of each other. Because the same information is contained in both the upper and the lower sidebands, sometimes only a single sideband is transmitted, and this is called **single sideband** transmission **(SSB).** Also, the carrier is sometimes suppressed because it, too, contains no information, and this is called **suppressed carrier** transmission **(SSC).**

bandwidth — the width of the band of frequencies that constitutes a specific signal or is passed by a communications channel or device. According to the French mathematician Fourier, any complex waveform can be decomposed into the sum of pure tones, or *sine waves,* at different frequencies and with appropriate maximum amplitudes and phases. A graphical plot of the frequencies that characterize the signal is called the frequency spectrum, or simply the **spectrum,** of the signal.

bps — the abbreviation for bits per second, the rate at which bits are transmitted. A bit represents a single *binary digit* and is either a zero or a one. *Digital signals* are represented by bits. The bandwidth required to pass a digital signal is approximately half the bit rate.

dB — the abbreviation for *decibel,* which is a measure for comparison of two quantities with one another. The decibel is the logarithm of the ratio of the two quantities.

digital — a process for representing an *analog signal* as a series of numbers, or digits, that represent its instantaneous amplitude variation. The process of converting an analog signal with a maximum frequency of F_{MAX} to a digital signal consists of sampling the analog signal in time at the **Nyquist sampling rate** of $2 \times F_{MAX}$ samples per second. These

---

sample values are then **quantized** into a fixed number of levels, usually chosen to be a power of two, i.e., 2^n. The quantized sample values are finally encoded as *binary* numbers utilizing *binary digits,* or *bits.* The digital process gives considerable immunity to noise, but this immunity is obtained at the expense of considerable bandwidth. The bandwidth required for the digital signal is $n \times F_{MAX}$.

diode — a device used in electrical circuits, which passes current flowing in one direction and blocks current flowing in the opposite direction. Diodes are usually constructed from a junction of semiconducting materials.

filter — an electrical circuit designed to pass or to block certain frequency components in a time-varying signal. A filter that passes high frequencies is called a **high-pass filter (HPF).** A filter that passes only low frequencies is called a **low-pass filter (LPF).**

frequency modulation (FM) — a technique for continuously changing, or *modulating,* the frequency of a high-frequency tone, or *sine wave,* called the **carrier,** in synchrony with the instantaneous amplitude variations of an information-bearing signal. One effect of frequency modulation is to translate and expand the frequency spectrum of the information-bearing signal. Another effect is to increase the *noise immunity* of the FM signal, but this immunity is obtained at the expense of a greater bandwidth required by the FM signal compared to amplitude modulation.

Hz — the abbreviation for **hertz,** which is the unit of measurement of the **frequency** of a time-varying signal. The frequency of a *periodic waveform* is the number of full repetitions, or *cycles,* that occurs in one second. For example, a pure tone that repeats a full cycle every 0.1 second would have a frequency of 10 Hz. High frequencies are usually written using Greek abbreviations. A frequency of 1000 Hz would be written as 1 kHz, where "k" stands for *kilo.* A frequency of 1,000,000 Hz would be written as 1 MHz, where "M" stands for *mega.* A frequency of 1,000,000,000 Hz would be written as 1 GHz, where "G" stands for *giga.*

impedance — a measure of the total opposition to the flow of an alternating current in an electrical circuit. The total opposition varies with the frequency of the current. Impedance is measured in the units of ohms.

ohm — the unit of measurement for electrical resistance and impedance of an electrical circuit or device, denoted by the Greek letter *omega* (Ω).

servomechanism — an electrical circuit used to compare one signal with another. One signal is a desired reference signal, and the second signal is a measure of the actual performance. The servomechanism will issue the appropriate instructions to control or adjust the second signal to minimize the difference between them. In this way, errors in desired performance can be corrected.

sine wave — a pure tone consisting of a single frequency. A sine wave has a maximum amplitude and a single frequency that characterize it. A sine wave is a smoothly time-varying waveform that alternates in *polarity,* or direction of flow (either positive or negative).

spectrum — a graphical plot of the frequencies that constitute a signal or are passed by a communication channel or device. The spectrum of a periodic signal consists of single frequencies only at integer, or *harmonic,* multiples of the **fundamental frequency.** The fundamental frequency of a waveform is the rate of repetition of the shortest repetitive full cycle of the waveform. Such a spectrum is called a **line spectrum,** since it consists of only lines at harmonics of the fundamental frequency. No signal can be perfectly periodic, and these sharp lines would usually be broadened somewhat.

time constant — a measure of the speed with which an electrical circuit responds to a fast-changing signal. The time constant is the amount of time required for a sudden change to decay to about 37% of its initial value.

transformer — a device consisting of two coils of wire that are closely coupled magnetically with each other. An ac signal flowing in the input, or **primary,** coil induces a similar signal in the output, or **secondary,** coil. The voltage amplitude of the output signal can be made larger or smaller than the input signal, depending on the ratio of the number of turns of the coils for the primary and the secondary.

vestigial AM — a form of amplitude modulation in which the upper sideband is transmitted in its entirety but only a small portion, or vestige, of the lower sideband is transmitted. This saves bandwidth compared to conventional **double sideband** transmission, and the demodulator at the receiver is still relatively simple. A more efficient method in terms of bandwidth would be **single sideband** transmission, but the demodulator is more complex.

volts — a measure of the electron-moving force in an electrical circuit, abbreviated V.

watts — a measure of the power of a signal or waveform, abbreviated W.

This glossary has been partially adapted from the one contained in *Introduction to Telephones and Telephone Systems,* A. Michael Noll (Artech House, 1986).

Index